John Bower

Simple methods for detecting food adulteration

John Bower

Simple methods for detecting food adulteration

ISBN/EAN: 9783337201388

Printed in Europe, USA, Canada, Australia, Japan

Cover: Foto ©berggeist007 / pixelio.de

More available books at **www.hansebooks.com**

SIMPLE METHODS

FOR DETECTING

FOOD ADULTERATION

BY

JOHN A. BOWER

AUTHOR OF "HOW TO MAKE COMMON THINGS," ETC.

"Apply the principles of science, and make them available to the needs, the comforts, and luxuries of life."—TYNDALL.

WITH 36 ILLUSTRATIONS

PUBLISHED UNDER THE DIRECTION OF THE GENERAL
LITERATURE AND EDUCATION COMMITTEE.

LONDON:
SOCIETY FOR PROMOTING CHRISTIAN KNOWLEDGE,
NORTHUMBERLAND AVENUE, W.C.; 43, QUEEN VICTORIA STREET, E.C.
BRIGHTON: 129, NORTH STREET,
NEW YORK: E. & J. B. YOUNG & CO.
1895

RICHARD CLAY & SONS, LIMITED,
LONDON & BUNGAY.

PREFACE

WE hope that this little book will be useful in more ways than one. It contains a series of samples for experiment, and another for observation, each of which by itself should be of educational value.

The experiments are simple, consisting for the most part of the application of chemical tests, the results of which must be carefully noted, and the comparing of weights of certain substances, generally called "taking their specific gravity." The observations bring in the use of the microscope; these will be better performed by a more elaborate than by a cheaper microscope. It is not, however, the privilege of everybody to have the use of an instrument of very high power. Each must, therefore, do his best with the instrument at his command.

The time may come when we shall have the use of a high class instrument at Technical Institutes or even at free libraries, where under proper supervision we shall be allowed to examine any specimens about which we may require information, and which we may ourselves supply. A simple microscope can be bought for a few shillings, while an elaborate instrument costs several pounds.

Very much, however, of this book can be done without the microscope, although much better with it. To carry out the experiments no expensive apparatus is required. In Appendix No. 2 will be found the approximate prices of the items mentioned in the book.

The whole forms a course of technical training.

In the second place, the exercises are confined to substances used in food, about which we certainly ought to have some information, for on the purity of food good health greatly depends. Without good food no family can enjoy sound health, and no food can be classed as good if it be adulterated.

From returns that we have carefully examined, we are glad to find that adulteration is decidedly on the decrease; but when cases turn up of such

extensive character as we frequently see reported
in our daily papers, we are forcibly reminded that
we are at any time liable to fall victims to this
fraud.

The directions in this book refer to only the
simplest methods of detecting adulterations.
Where substances are troublesome of detection,
or where great accuracy in a common case is
required, it must be left to a professional analyst.

Milk is one of the important articles of food
that suffers most from adulteration, for hardly a
week passes without some prosecution for adding
water to it or taking cream from it. Fortunately
the method of detecting either of these frauds is
a simple one, not that a buyer can prosecute, but
in confirming his suspicions by an experiment, he
can make a complaint to the vendor, and if the evil
is not remedied he can call the attention of the
sanitary inspector, who will take up the case.

Genuine tea and coffee can by very simple
methods be distinguished from adulterated samples,
and the same remark applies to many other
articles of food.

We are greatly obliged to several scientific
friends who have been good enough to read
through the proof-sheets of this work, and one of

these friends says that "the book is a terrible indictment of modern civilization." We hope, therefore, that it will be useful in helping to improve the aspect of things, by reducing still further this evil of adulteration, which we should like to see entirely stamped out.

We must also tender our best thanks to Mr. Horace F. Taylor, who made for us the careful drawings by which this little work is illustrated.

JOHN A. BOWER.

June 1895.

CONTENTS

LIST OF ILLUSTRATIONS

SIMPLE METHODS FOR DETECTING FOOD ADULTERATION

CHAPTER I

INTRODUCTION

THIS little book is not written with the idea of alarming the householder, but rather to put into his own hands the methods of detecting *ordinary* adulterations by very simple means—means that should always be at hand in the *ordinary* household.

We hear it sometimes said, " Oh, I don't believe in this adulteration—our food is not adulterated." Fortunately it is not adulterated to an alarming extent; that is, however, not the point; it should not be adulterated at all.

If you buy milk, are you to be supplied with milk and water ? You ought to have milk only— pure milk; if you wish to add water you can do

so without paying fourpence or fivepence a quart
for it.   If you buy tea, are you to have a mixture
consisting largely of spent leaves?   Again, if you
want coffee, are you to have one-quarter or more of
it made up with chicory? and butter, is it to be
simply butter, or is it to be butter mixed with
other fats, sometimes indeed consisting principally
of " other fats " ?   Margarine and substances used
in adulterating butter are all very good in their
places, but not as substitutes for butter; nor has
chicory the same food value as coffee; and there
is no goodness to be extracted from spent tea-
leaves.

As we have mentioned, adulteration is not
now going on to an "alarming" extent, but the
Summary of Reports of Analysts showed that for
the year ending 1891, about 12 per cent. of *all
the food sold in this country* was adulterated, and
that the articles that suffered most were spirits,
coffee, tea, and butter.   Since that date, however,
adulteration has been on the decrease, so that for
1895 the percentage is not so high as that for
1891.

As coffee comes second on the list, there is
no doubt but it is adulterated to a large extent;
sometimes, we are told, it amounts to 75 per cent.
In the face of such a statement as this, a little
guide which helps those who follow it to detect
adulterations cannot be very much out of
place.

Since the passing of the Adulteration Act, cases

of wilful adulteration have been much fewer; but even now, hardly a week passes without the exposure of some tradesmen or manufacturers, who have been pounced upon by inspectors, who hand the articles over to the public analysts; and when the cases are " gone into " in the public court, we are astounded at the methods adopted, some of them most elaborate, some most flagrant, but all showing plainly enough that the public are " being taken in."

Now it is arrowroot, largely adulterated with various starches; then it is milk with 20 per cent. or 30 per cent. of water; then it is the bread; then it is olive oil, that has not a particle of real olive oil in it; then it is the mustard, coloured with turmeric and bulked with wheaten flour; and so we might go the round of substances we buy as pure food day by day.

Now let us refer to the Act, so as to get the full meaning of the term "adulteration." It says, "The article shall be held to be adulterated—

" 1. If any substance or substances has or have been mixed with it so as to reduce, or lower, or injuriously affect its quality, strength, purity, or true value.

" 2. If any inferior or cheaper substance or substances has or have been substituted wholly or in part for the article.

" 3. If any valuable constituent of the article has been wholly or in part abstracted.

" 4. If it be an imitation of, or be sold under the name of another article.

" 5. If it consists wholly or in part of a diseased, or decomposed, or putrid, or rotten animal, or vegetable substance, whether manufactured or not; or in the case of milk, if it is the product of a diseased animal.

" 6. If it be coloured, or coated, or polished, or powdered, whereby damage is concealed, or it is made to appear better than it really is, or of greater value.

" 7. If it contains any added poisonous ingredient which may render such an article injurious to the health of the person consuming it."

The following sections of the Act show its real working powers :—

"Section 6. No person shall sell, to the prejudice of the purchaser, any article of food or any drug which is not of the nature, substance, and quality of the article demanded by such purchaser, under a penalty not exceeding £20, provided that an offence shall be deemed to be committed under the sections following.

"Section 7. No person shall sell any compound article of food which is not composed of ingredients in accordance with the demand of the purchaser; penalty not exceeding £20.

"Section 9. No person shall with intent sell to a purchaser a substance in its altered state without notice, nor abstract from any article of food any part of it, so as to affect injuriously its quality, substance, or nature, and no person shall sell any article so altered without making a disclosure of the alteration, under a penalty in each case not exceeding £20."

By the clauses of this Act, it is seen that the tradesman is fined on conviction, but in France the adulterator is put into prison, and the authorities shut up his shop so long as he is in prison; the shop has a label put on to it, stating that it is shut up because the proprietor is convicted of adulteration, and that the proprietor himself is in prison. We fancy that such a punishment as this once inflicted, prevents most effectually the repetition of the folly in the same individual.

While this Adulteration Act secures to us as far as possible that all the necessaries of life shall be supplied to us as pure, we on our parts should see that we get them in this state of purity. We should do nothing to encourage tampering of any sort, by offering to take damaged goods at a lower price, or by beating down a tradesman from a fair price. If goods are offered at an unusually low price, something wrong with them may be suspected.

One other source of adulteration ought to be noticed, viz. that arising from carelessness. Many articles of food are exposed to the air, sunlight, and to any amount of dust that may accumulate on them, and then these articles are sold as pure, and at the same price as fresh goods. Take sugar, rice, currants, and raisins, which are

B

generally exposed in a tradesman's window; they cannot remain there many days without getting covered with dust and dirt, which neither improve their appearance nor their nutritive value. This ought not to be; articles of food should always be carefully preserved from any source that may injure them.

The most abominable cases of adulteration are, however, those where inferior and damaged articles are systematically bought, glossed over and made up, and sold to the public as pure and of good quality, at sometimes a trifle less than is given for the best goods.

In the methods for detection of adulteration we shall keep to quite the simplest, such as the application of an ordinary chemical test, by weighing, and by the use of the microscope. As almost everybody learns chemistry now-a-days, the application of the simple tests will be interesting as well as instructive. At the same time, it will require care and discrimination, it will bring out the powers of observation, making the experiment itself a very useful exercise. We are not supposing that you are to test every fresh·loaf of bread that comes into the house, or every pound of tea, coffee, or sugar, but that you should select samples occasionally, and especially those that give rise to suspicion.

The next process we recommend is one·of weighing; for that purpose you should secure a thin stoppered bottle, that shall hold exactly 250 grains of distilled water, or if it is not exactly filled by that quantity, the bottle should be marked at that point: you can then take the specific gravity of milk, butter, oil, and such things easily. You must also have a counterpoise for the bottle,—this can be managed with small shot,—then whatever may be the weight of the substance tested, multiply it by 4; that will be its weight compared with water at 1000. A small set of apothecaries' scales and weights will be all that is necessary to carry out these experiments with fair accuracy.

Then, lastly, there is the microscope; in this you will take great interest, for it will help you to examine many things outside the region of food. A low power lens will do for the examination of large objects, such as sections of tea-leaves, but for the starches you will require a much higher power.

The substances to be examined are to be placed on a glass slide supplied with the instrument, to be covered with a drop of water, or with a little glycerine and water as the case may be, then covered with a very thin clear piece of glass, supplied also with the instrument. The mirror under the stage of the instrument is to be so placed that it will throw light on the object to be examined. Of course it is quite necessary for you to know what you have to look for in each case; we shall therefore give such particulars as we go along. In Fig. 1, the diagram shows how the double convex lens enlarges the image of an object.

We hope that we shall be able to present a series of interesting exercises, even if adulterations have not to be detected. We propose to give directions for preparing such of the tests as we apply; at the same time, we should advise you to procure the series neatly done up together in a case, so that they are always at hand.

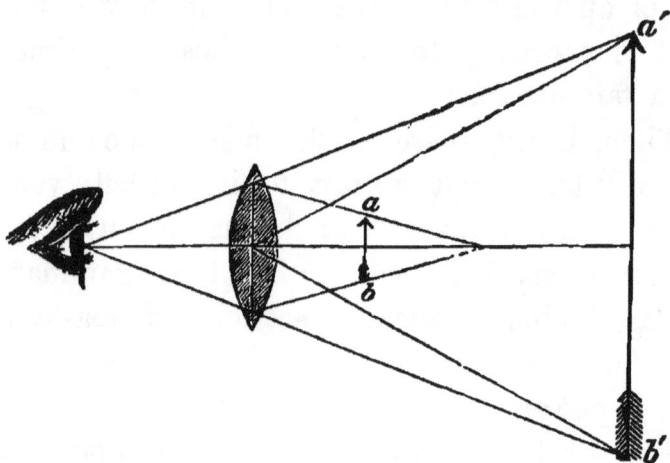

Fig. 1.—Magnifying power of Convex Lens.

Where the methods are difficult we shall omit them, as they are only suited to the professional chemist. You must not therefore look upon this as a thorough series of tests for adulterations, but only a handy-book giving the simplest possible methods of detecting the most glaring and the most easily discovered.

We do not advocate a general raid on our tradesmen, for most of them are not even to be suspected of adulterating food ; but if some of them will insist on carrying on this fraud, we have no business to allow their false and fictitious articles to be passed off upon us, because it is condoning dishonesty and encouraging trade immorality.

CHAPTER II

BREAD, FLOUR, AND GRAIN FOODS

BREAD is often called the "staff of life." Till lately, however, it was greatly adulterated, carelessly made, and certainly had no claim to the above title. Excellent bread is now made by most of our bakers, and since the Bakers' Exhibitions of the last two years, many improvements in mixing and baking have been brought into practice. Some of our bakers, however, think they add to their repute by attaching to their advertisements the results of certain analyses of samples of their bread submitted to an analyst. Such a certificate is of no value, for the published result only belongs to the one specimen referred to. A good and reliable baker needs no "puff" of this kind to secure his success. Let his bread be always good and pure, and his customers will appreciate him, and support him accordingly. A loaf of good bread can be told by the smell, uniformity of its pores, its lightness, thinness of its crust, and that crust baked to a nice brown. If when cut, a loaf at a day old has a disagreeable sourness about it, that is not a good sample.

The general adulterations are limited to a little alum, a little too much salt, and more potatoes than are needed for the leaven.

Of these adulterations alum is the worst; it enables an inferior flour to be made up so as to look as white as a better flour; it also enables the flour to take up more water. Too much potato imparts a crumbleness to bread, and too much salt appeals to the taste. Alum makes bread less digestible. This is a great reason why it should not be used, especially as food for young people. Some bakers will tell you they cannot make bread without alum; that is not true. Persons who make their own bread at home do not use alum, and with a little practice bread may be very successfully made at home;—we wish it were a more universal custom than it is. Once accustomed to home-made bread, bakers' bread would very rarely be used again.

Now for a ready means of detecting alum. Cut a small square out of the crumb of a loaf, put it in a plate, pour on it a mixture of a tincture of logwood and carbonate of ammonia solution—the mixture to consist of equal quantities of each. If no alum is present in the bread the pink colour will remain, but the presence of alum turns it blue. We have never found this test to fail, although we have used it for the detection of alum for the last twenty years. To ascertain the quantity of alum is a more difficult matter, and one that requires more chemical skill than we can here direct. The detection of potato starch and wheat starch after they are made up into bread is also one that we cannot detail to advantage. They may be examined, however, by the microscope. The logwood tincture is easily made, by steeping some chips of logwood in methylated spirit, $\frac{1}{4}$ oz. of the chips to 1 oz. of spirit. Keep it closely stoppered. The carbonate of ammonia is to be a saturated solution. In using this test take equal parts of each, and add three times its volume of water before putting it on the bread. This solution will detect as small a quantity as four grains to four pounds of bread, which is the weight of a quartern loaf.

Sometimes thin strips of gelatine are stained with the logwood and ammonia mixtures, and applied to the bread to be tested with equally satisfactory results. Should any other earthy salts be present, like magnesia, the tint will be a higher purple still. Alum is the only substance you are likely to find.

The material containing the alum is sold to the baker as "hards" and "stuff," and this is really a mixture of alum and salt. It is kept in bags, weighing from a quarter to 1 cwt. The very worst flours are the most assisted by the use of alum, so that even damp and mouldy flour can be used, and made up into white bread.

Good new bread should contain about 45 per

cent. of water, but we are informed that some bakers manage so that the quantity of water retained is much larger. They manage this by putting the dough into an over-heated oven. The outside of the bread is rapidly browned, while the inside is not properly cooked. Be therefore cautious in buying "slack-baked" or "under-baked" bread.

Well-made and well-baked bread retains this normal amount of water for several days, so that the dryness noted in a stale loaf is not due to loss of water.

Should carbonate of ammonia be used in bread-making—it is for whitening and raising it. If the bread is very white it may be suspected. It may be detected by mashing a slice of bread in cold water that has been boiled for some time, or in distilled water. After the soaked bread has stood for half-an-hour or so, strain off the water, add a few drops of hydrochloric acid, evaporate the liquid to dryness, then add some strong caustic potash solution. If ammonia is present you will at once be made aware of it by its pungent smell.

To evaporate such liquids as these the temperature must not exceed that of boiling water. This can be done by using a small saucepan, in which a saucer containing the liquid to be evaporated takes the place of the lid. The saucepan must be kept boiling till the evaporation of the liquid in the saucer is complete, and a little space must be

left between the rim of the saucepan and the edge
of the saucer for the steam to get away.

Biscuits and cakes are made with mixed substances, so that it is difficult to select which of the
starches would be an adulteration. When colouring substances are used, to make up for the
absence of eggs, that colour is an adulterant.
The colour itself may be harmless, but it is never-
theless an adulterant. If you wish, however, to
convince yourself that colouring matter has been
used, dry, and then crush up some of the crumbly
portion, and make a solution of it, which will be
coloured according to the quantity that has been
employed.

The larger firms of biscuit manufacturers may
be relied on as using the best and purest materials
only, and nothing that is foreign to the substances
actually required.

Flour.—When we talk of flour, we mean
wheaten flour pure and simple, and made of
grain, free from damage of any kind. Damaged
wheats are unfit for food, and all the dressing in
the world cannot make good flour out of them.

Wheaten flour consists of starch and gluten;
the latter substance gives to it its nutritive value,
and is contained in the outer skin of wheat, while
the starch is a heat-giving substance, and forms
the inner part of the grain. The gluten gives the
stiffness to dough, and enables it to rise in bubbles,
and become light, as it is called. Whole grain is
used for making brown bread. Coarse brown

bread is not so easily digested as white bread, or
as that made from whole grain.

To examine flour, you must call in the aid of
the microscope. You must first separate the
starch from the gluten, by placing a pinch of
flour on a piece of muslin or calico tied loosely
over the top of a glass. Let water trickle
through it, and you can knead the flour at the

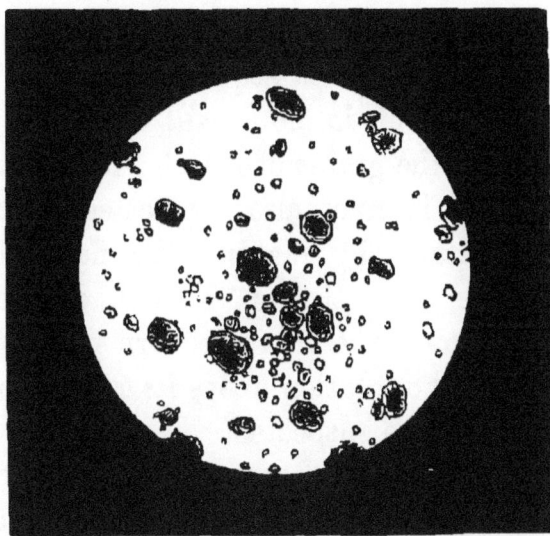

Fig. 2.—Granules of Wheat Starch.

same time with the fingers. The starch will pass
through the cloth, and form a layer at the bottom
of the glass. Let it settle, then pour off the
water; put a few grains of this on a glass plate
for examination. Place a drop of the starch
solution on a slide, and cover with the very thin
glass slip. In Fig. 2 we give the shape of wheat
starch granules; if you have anything besides these,

it is due to adulteration. We give illustrations of the other starches; a little practice will soon enable you to distinguish one from the other. Microscopic slides can be bought with the various starches; they cost but little, and to learn from these as standards you will soon be able to recognize the different kinds as easily as letters in the alphabet. A magnifying power of 250 diameters should be used. As no two starches are alike in form, such an acquaintance with them as we recommend is a great acquisition.

Alum, if present in flour, will show itself distinctly among the grains of starch; if its presence be suspected, it may also be tested with the logwood mixture, by making the flour into a creamy paste, and stirring in with it the test solution. Should the mixture turn purple alum is present; if the solution keeps its colour, nothing of the sort may be suspected.

Millers tell us that sometimes alum gets in from the mill-stone dressings, and in the case of the hard Egyptian wheat, alumina does get in from clay which fastens itself in minute quantities to the outer shell, because the threshing is done on clay floors.

When flour is adulterated with damaged wheat it may be detected without the microscope, for if a little solution of aniline violet be used with the flour, the damaged starch takes up the colour at once, while the sound starch refuses to do so.

If rice starch is used, the microscope will at

once detect it; the form of the granule of rice is

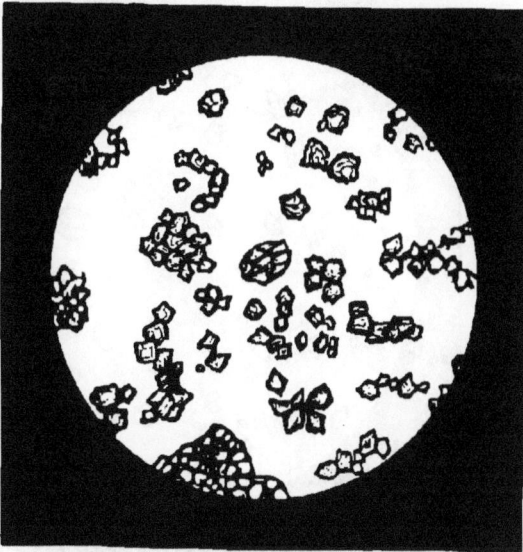

Fig. 3.—Granules of Rice Starch.

Fig. 4.—Granules of Indian Corn.

square, as seen in Fig. 3, and very much smaller than that of wheat.

If Indian corn be used in adulteration, it is

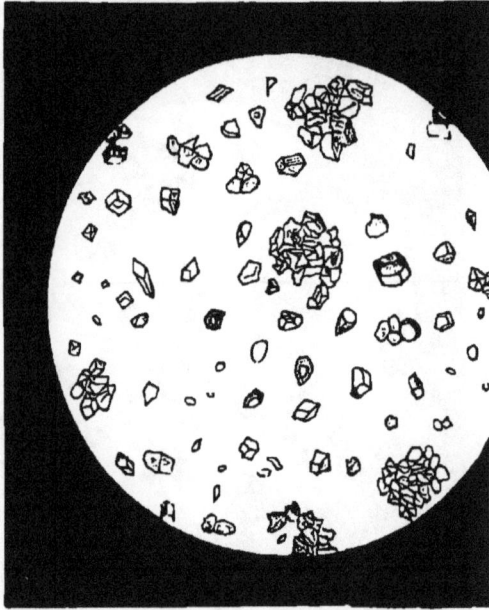

Fig. 5.—Oat Starch.

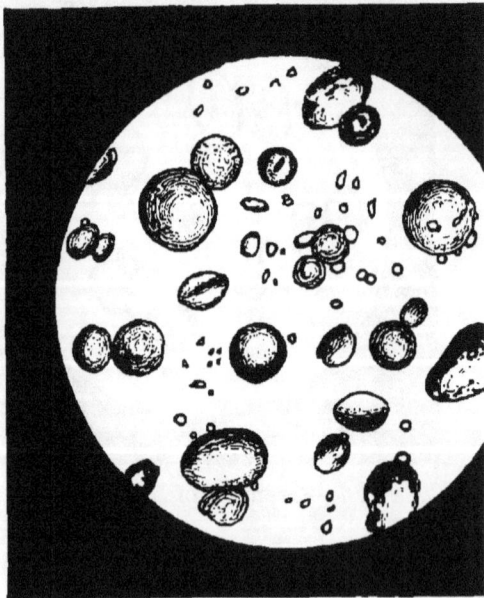

Fig. 6.—Barley Starch.

easily detected by the microscope; it is more irregular in shape, and is much larger. In Fig. 4 we give the microscopic appearance of Indian corn starch. Indian corn is richer in fat, and tends to make wheaten flour into a softer dough if mixed with it, and gives bread made with this mixture a peculiar sweetness.

In some corn-flours that are sold Indian corn or maize is the sole ingredient.

Oatmeal.—This cannot be made very well into bread, but is used largely for porridge and cakes. It is as porridge a most nutritious food for children; it, however, requires well boiling. Oats are also used for making Emden groats. The adulteration to which it is most liable, and to which it suffers frequently to the amount of 15 per cent., is barley-meal. As the starches of the two are, however, so different, the microscope will soon detect it. Take a little of the mixture, put it on the slip of glass, drop on a little water, cover the thin glass over it, and put it on the stage of the microscope and examine it. Should the two starches be present, count the number of granules of each, you will then at the same time get the proportion of the adulterant and the form of the granules. In Figs. 5 and 6 we give the starch granules of oats and barley. A second or even a third slide should be prepared and examined in like manner. You can then take the average result from the three examinations.

CHAPTER III

OTHER STARCHES USED IN FOOD

Barley.—THIS grain is sometimes freed from its husk, so as to be used for broths and barley-water. In this condition it is called pearl-barley. Mixed with flour, as we have mentioned, it makes an inferior bread. To detect its presence use the microscope; we referred to the form of its starch granule in the last chapter (Fig. 6). The principal use of barley is in the manufacture of malt. From the malt an extract is made, under the name of *maltine*. This, when prepared at a low temperature, has the power of assisting very much the digestion of starchy food.

Under this heading we can deal with the sundry "farinaceous foods" made up as fit for infants. None of them are so good as meat broths or albuminous foods, for the infant digestion is not fitted to digest starch foods.

To detect adulterations, let a portion be dried, by the method we recommended, over a saucepan of boiling water, and when dried it should be examined by the microscope. Compare the granules of starch with potato, and those of the other figures given. Should potato starch be present in larger quantities than other starches, it should not be used as infant food, for this form of starch is more difficult of digestion than any other.

Rye.—Rye bread is not a favourite in this country, although it is used on the continent.

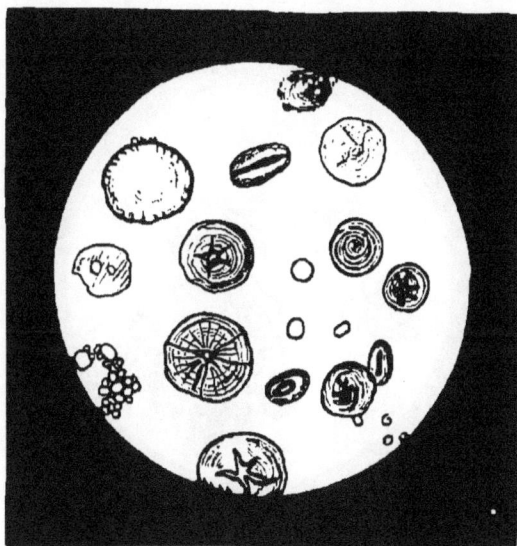

Fig. 7.—Rye Starch.

We give the form of its starch granule in Fig. 7, because it is used to adulterate various starch foods used in this country. The flour from rye contains a gluten, which brings rye-bread nearer to a wheaten bread than that produced from other grain. Rye makes a very dark bread, and it is heavy and poor in flavour.

Indian corn.—We referred to this substance as

c

used occasionally to adulterate wheaten flour, and gave the form of its starch granule in Fig. 4. By itself it cannot be made into bread, because it contains no gluten. It is, however, a favourite substance for making cakes in America, and these are very nutritious. In this country it is sold as corn-flour and as hominy, both of which are prepared from maize.

Fig. 8.—Granules of Tapioca.

Rice, tapioca, and sago.—These are all forms of starch food, and are used largely in this country, but in the form of puddings and cakes, mixed with other ingredients which assist in their digestion. In Fig. 3 we gave the form of the granule of the rice grain; once recognized it is not likely to be forgotten. It is small and angular in shape. Mixed with wheaten flour, to which, as we have said, it is sometimes added for the sake of whiteness, it can readily be distinguished.

By itself rice is the poorest sort of food; in using it for puddings it should be well boiled. Ground rice is not easier of digestion than the whole grain. If bought in the grain, there is not much fear of its being adulterated.

In Figs. 8 and 9 we give the forms of tapioca

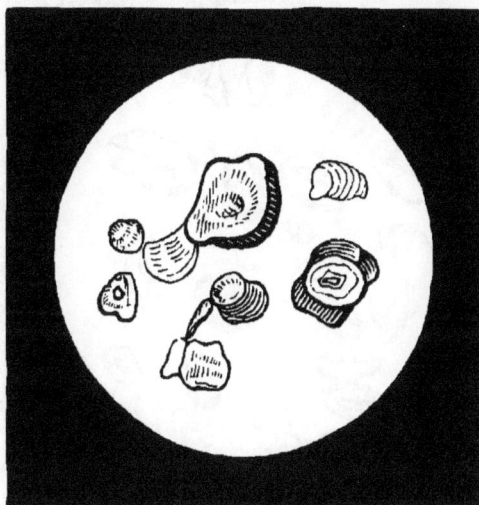

Fig. 9.—Granules of Sago Starch.

and sago starch granules, the latter elongated, rounded at one end, truncated at the other Unlike other starches, tapioca starch is rendered gummy by the action of ammonia.

Maccaroni and vermicelli.—These substances are manufactured from hard wheat, and being rich in gluten, are very nutritious. To detect adulteration with other starches, some of the material must be powdered and washed, that the starch portion may be examined under the microscope.

Peas.—Besides using peas as a vegetable in its green state, we use them when fully ripe in their dried state. In this condition it is true they contain a large amount of nourishment, but require considerable cooking to render them digestible.

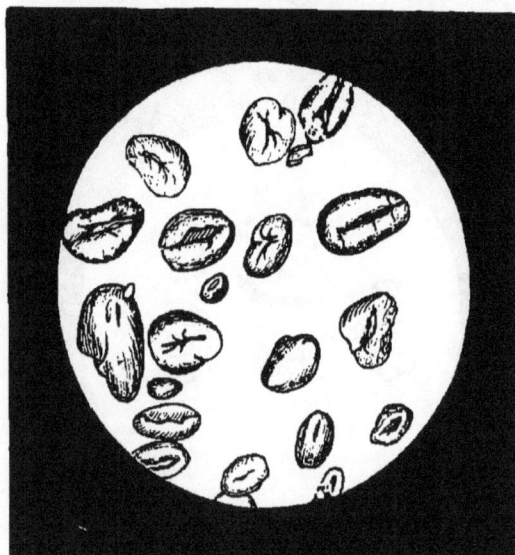

Fig. 10.—Pea Starch.

As green peas they are when cooked easily digested and very nutritious.

In the dried state peas are sold as split peas and as pea-flour. Pea-flour is very rarely adulterated. In Fig. 10 we give the form of its starch granule, so that under the microscope it is easy to tell whether you have pea-flour or the starch of any other grain.

Beans.—As a food the bean is of about the

same value as the pea. It is very rarely sold to the public as a bean-flour, if so, it is only used to adulterate other flours. We therefore give the form of its starch granule in Fig. 11. By comparing Figs. 10 and 11 the presence of each can be readily detected.

It is most convenient that these starches all

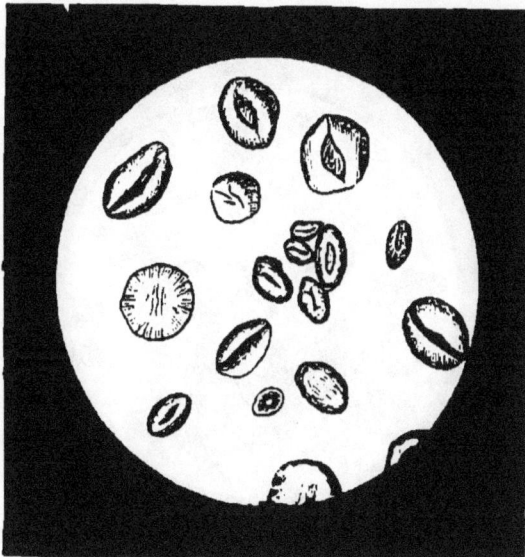

Fig. 11.—Bean Starch.

differ in shape, size, and surface, so that when viewed under the microscope they may be so readily distinguished from one another. The microscope really becomes the most certain guide in all inquiries about the kind of starch present in any special kind of food.

Arrowroot.—No starch food is perhaps so important to the invalid as arrowroot. This is not

on account of the nutriment it contains, but it is
so easily digested. It is a food which may be said
to consist of starch only. It is also one of those
which lends itself to be greatly adulterated. These
adulterants are starches of 'other kinds, some of
which we have been dealing with already. That,
however, to which we have only briefly referred is
the very starch that is the most largely used for

Fig. 12.—Potato Starch.

this purpose of adulteration : that is potato starch ;
its starch granule we give in Fig. 12. You see
how very widely it differs in shape and character
from the starches we have already sketched. The
granules vary very much in size and shape—some
are nearly round, while others are like flattened
oyster-shells. The larger granules have distinct
concentric rings surrounding a prominent eye, in
others this eye is a mere spot at its narrowest end.

There is, however, no fear of being unable to distinguish potato starch.

That sold as British arrowroot consists very largely of potato starch. Now let us see what real arrowroot should be.

We have what is called the East Indian and West Indian arrowroots, and also a preparation called Brazilian arrowroot, which is really a preparation of tapioca. Then there is the Florida arrowroot. The West Indian is the best, the most esteemed being that which is grown in Bermuda; next in reputation is that grown in Jamaica; some merchants tell us it is quite equal to it.

Arrowroot is obtained from a species of maranta. The roots of the plants are often more than a foot long, almost as thick as a finger, and are covered with large white papery-looking scales. When the plants are about a year old these roots are dug up, peeled, and pulped by machinery. The pulp is mixed with water, which dissolves all but the starch; this settles to the bottom of the vessel in which the process is performed. It undergoes several washings, and is finally dried in the sun or in drying-houses. Dust and insects are shut out by gauze protectors. A genuine maranta arrowroot consists of nothing but this dry, light, white, opaque starch, a sketch of which granules we give in Fig. 13, and of an inferior East Indian quality in Fig. 14. The latter kind is often used to adulterate the former. It comes over to this country in very carefully packed cases, and if you get the best quality it should feel dry, and produce a slight crackling noise when rubbed between the fingers. If you buy a cheap quality of arrowroot, you may be sure it is inferior, for it is frequently sold in this country at a less price than is paid for the

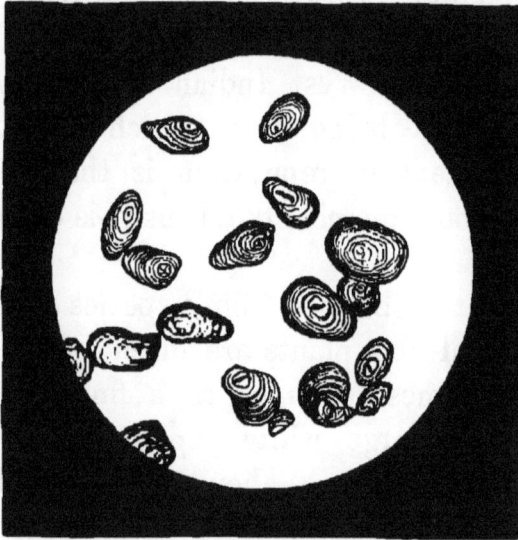

Fig. 13.—Maranta Arrowrcot.

genuine in the country that produces it. The low-priced articles, as we have said, contain very little genuine arrowroot, but large quantities of potato starch. From this you will see that although a starch substance is mixed with starch, it makes all the difference in the world as to its quality as a food for invalids. Arrowroot starch is easily digested, and therefore makes a good food for invalids; while potato starch is not easily

digested, and should therefore be withheld from
invalids and children. Sago-flour, rice-flour, and
even wheaten starch are sometimes used as adulterants. We have, however, so fully described the
characters of these starches, that when you put a
sample under the microscope for examination, you
will very readily be able to see how much of the

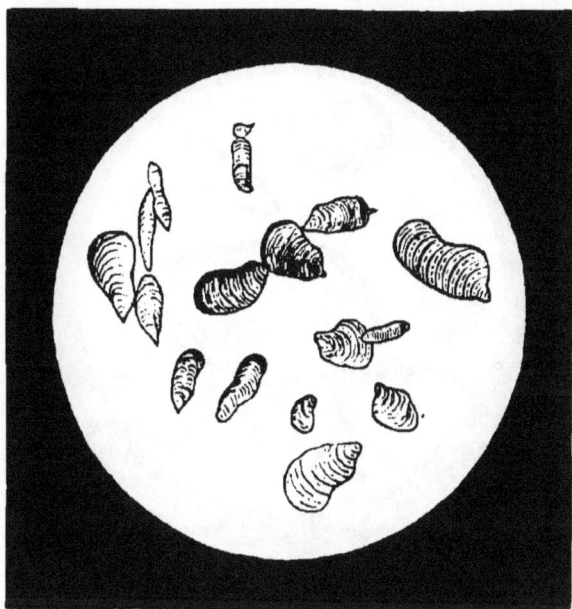

Fig. 14.—E. Indian Arrowroot.

genuine article you have really got. In studying
Figs. 13 and 14 you get the distinctive forms of
the arrowroot. They are convex, more or less
oval, although somewhat irregular, and they do
not differ very greatly in size, and are surrounded
with delicate concentric rings.

Dry arrowroot is without smell, but when

dissolved in boiling water it seems to acquire a
peculiar smell, while it rapidly swells up into a
perfect jelly.   In Fig. 15 we give the form of the
largest starch granule as it exists in Canna arrowroot, or *Tous-le-mois.*

There is also a special chemical test which can
readily be applied to detect these two starches.

Fig. 15.—Canna Arrowroot.

Under a drop of sulphuric acid the granules of
potato starch take up a most beautiful reticulated
appearance, resembling fir-cones, which gradu-
ally softens down as the starch becomes gelatinized.
Upon maranta starch, sulphuric acid produces no
such change, and it keeps its form for some time
before it gelatinizes. We have not taken into
account here that damaged samples of arrowroot

are sometimes used in adulteration. This will reveal itself, however, under the microscope. Should there be mould or animalculæ of any kind present, you may be sure you have an old and damaged sample under examination. This should not under any circumstances be used for food.

CHAPTER IV

MILK, CREAM, BUTTER, AND CHEESE

Milk.—Good milk should be opaque, white, but have no solids floating in it, and be of a sweet, agreeable taste.

Perhaps no food substance sold in large quantities is so much adulterated as milk. The great adulterant is water. It also suffers by having a portion of its cream taken away before it reaches the purchaser. Chalk and other substances said to be used in adulteration are so rarely met with that we need not trouble ourselves about them. We have never discovered chalk among the samples we have examined. The adulteration with water, however, is a very serious matter, although it seems so simple.

In London alone, very nearly £1,500,000 is spent yearly for milk ; and according to statements which are official, the average adulteration with water extends to one-twentieth of the whole, which means that the cost of the water sold as milk in London alone, amounts to between £70,000 and £80,000.

Milk is especially the food of the young, and contains everything they want in the way of food —it is a type of a perfect food ; but added water robs milk of its value in this respect, according to the amount added.

Cow's milk contains solids amounting to one-seventh of its whole weight, of which one-third is sugar, a little less than one-third is cheese, and one-quarter milk fat.   The total solids in pure milk vary from 10·33 to 15·83 per cent.   This arises from circumstances we cannot discuss here; we mention it to show how difficult it is to state accurately the extent to which some samples have been watered.

Another mischief which is too often overlooked is, that the water used for adulteration is frequently foul, and may contain germs of disease.   The outbreak of several epidemics in various localities has indeed been traced to this source.   This danger is referred to by a writer in these terms: " Worse than the process of watering down, which does not merely impoverish good milk, it may happen that the water itself is impure, and a bucketful of foul well-water from a distant farm may mean a hundred deaths in one London parish."   To avoid mischief from this last form of adulteration, milk should always be boiled before being used.

As a food, cooked milk possesses only the same value as uncooked.

To be perfectly sure of the extent to which milk is adulterated, chemical analysis is necessary, but the following tests are sufficiently trustworthy for a general estimation.

First, by taking its weight compared with water, *i. e.* its specific gravity.

Supposing a quantity of water is taken which weighs 1000 grains, that same quantity of milk should weigh from 1029 to 1032 grains. This test can be carried out with the bottle we have already mentioned, which contains 250 grains of distilled water, for whatever the same bottle will weigh when filled with milk, will be due to the milk. Supposing we find that the bottle when filled with milk weighs 7 grains more than when filled with distilled water, that will be 257 grains; multiply this by 4, and we have 1028, *i. e.* a specific gravity of 1·028.

Another method of taking the specific gravity of a liquid, is to note how far a floating body sinks in it. This is the principle on which the lacto-meter (Fig. 16) is made. It sinks to a certain depth in average milk; if the milk is watered simply, it sinks to a greater depth; and if anything is added to increase the specific gravity, it does not sink to so great a depth. The instrument can be bought for about a shilling, and it is marked to sink to a certain point in good average milk: should the milk be of extra good quality, it does not sink so far; if of a poorer quality through

added water, it sinks to a lower level, and this
depth corresponds to the quantity of water
added to the milk, provided it has not been
robbed of its cream.   The scale marked on
the stem of the instrument enables you to

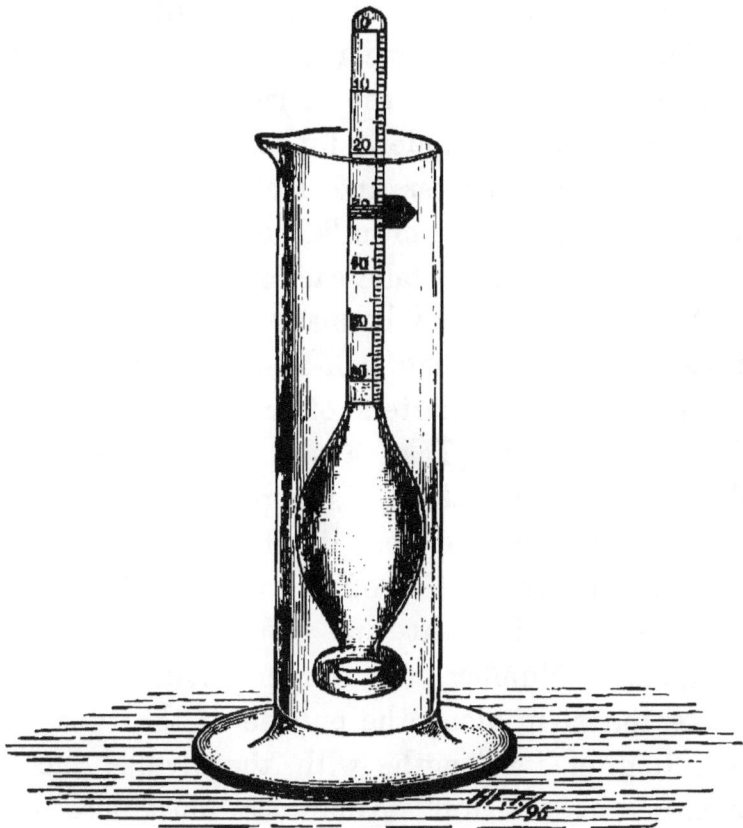

Fig. 16.—The Lactometer.

compare the quality of the milk with the depth
to which the instrument sinks.   It happens, however, sometimes, that milk from which cream has
been removed is adulterated with milk of a
similar quality—that is, skim-milk is added instead

of water; that increases the specific gravity, although the milk is poorer, for the skim-milk supplies it with solids, not fat, above its natural quantity. Owing to this, the lactometer cannot be looked upon as so reliable an instrument as the creamometer, which we will now describe.

The creamometer takes into account the amount of cream which rises to the surface when the milk is allowed to stand quietly for some time. When both the creamometer and lactometer are used with the same sample of milk, we get more reliable results than when we take each separately.

A creamometer may be made from an ordinary test-tube, not less than half-an-inch in diameter, better still three-quarters of an inch, and about five inches long. Take a slip of paper—a slip from the edge of a sheet of stamps is best, because it is already gummed; cut it off five inches long. With a pencil, mark this off in five equal divisions, and mark an inch at one end into ten equal divisions. Number the larger divisions, then paste the strip along the outside of the tube, in direction of its length, with the divided inch-measure at the top. Allowance must also be made for the rounded bottom of the tube. Fix this tube upright into a large cork or bung, and you have such an appliance as shown in **Fig. 17.**

To use it, pour into the tube sufficient milk to fill it to the height of the top edge of the paper slip; stand it aside in a cool place for six hours. At the end of that time read off the divisions occupied by the cream that has risen to the surface. Double that number, and you have the proportion in 100 parts, or the percentage of cream. If your sample of milk is good and pure, you will have from twelve to fourteen parts occupied by the cream ; if a poor watered specimen, it may be only half that number ; should it fall below ten per cent., you may be tolerably certain

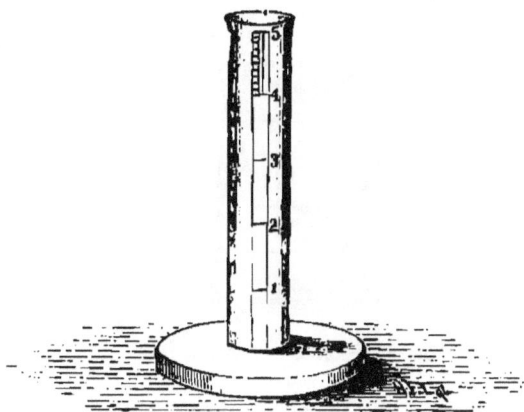

Fig. 17.—Test-tube fitted as a Creamometer.

that water has been added, or that cream has been taken away. The quality of the milk may be fairly estimated according to the number of divisions occupied by the cream. A watered or a robbed sample throws up much less cream in a given time than a pure sample.

Tubes on feet, with divisions marked on them, are made for this purpose. One may be bought for a small sum ; while within a few months a "tell-tale" milk-jug on this principle has been patented and put into the market. On

D

this jug, markings are given for the pint, half-pint, and quarter-pint; on handing this jug to the milkman for your supply, if he gives short measure it is at once detected. The estimate of cream is taken in the same manner as in the tube method, three scales being marked on it, for average, good, and very good milk. A sketch of this jug, which is sure to be a favourite with all but the milkman, we give in Fig. 18.

Fig. 18.—Tell-tale Milk-jug.

This method of testing the quality of milk is much more reliable than with the lactometer alone, owing, as we have said, to a form of adulterating new milk with skimmed or separated milk. As the skimmed milk has been merely deprived of its fat, it will still contain the other solids which are not fat. A milk so doctored will show a high specific gravity, although the yield of cream will be small. Curiously enough, it was by such an instance as this that the fraud of adding skim-milk was discovered.

It is only by a thorough analysis that the purity of milk can be fully certified, but this ready method of cream raising gives a general and fairly accurate estimate as to whether it has been adulterated or not.

The microscopic examination of milk is interesting. The quantity of fat particles in a good rich milk, contrast most favourably with the sparsely present particles in a watered or skimmed sample. In Fig. 19 we give a microscopic ap-

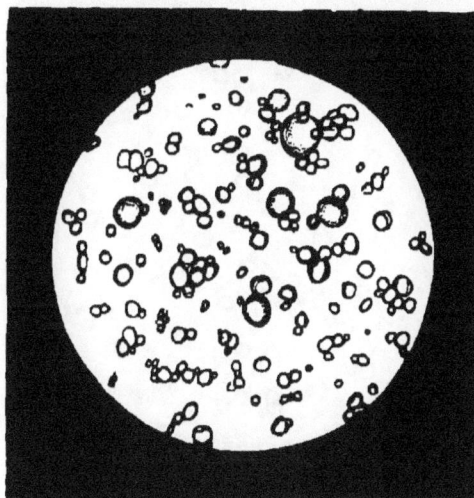

Fig. 19.—New Milk.

pearance of a sample of new milk; and in Fig. 20 a sketch of skimmed milk. The temptation to water a sample is so great, and so easily done, that the seller often does it without thinking of the consequences.

When milk is taken, it should not be stood near to any strongly-scented substances, for it so readily absorbs the sweet as well as the evil scents. In applying any of the above tests, the milk should always be thoroughly mixed, so that your sample for experiment is neither from the top nor from the bottom of your jug.

When it is impossible to get fresh milk, condensed milk may be used. Several good brands of this article are in the market. These tinned samples consist of milk reduced by evaporation to

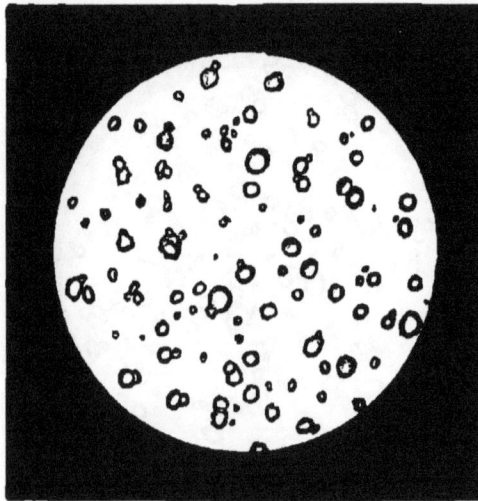

Fig. 20.—Skimmed Milk.

about one-seventh of its original bulk, and to this a little sugar is added to keep it good.

The greatest care should be used in selecting water to add to it; it should be of the purest quality, quite above any suspicion of contamination. Some persons prefer the sterilized milk on this account, for in it there is no chance of any germ of disease, such as may arise from impure

water. When condensed milk is mixed, it is well
to test it occasionally for the amount of cream it
will throw up.

Cream.—Experienced dairymen tell us that
cream may be thin and yet rich in fat, so it may
be thick, and yet not contain more than an average
amount of it. The cream that rises first to the
surface of milk is generally thin in quality, yet
it contains really more fat.

When you take the cream off the milk that you
have yourself set for raising, you are tolerably
sure of its purity; it is in the bought samples
that adulteration comes in. The colour should be
noted; it should not be white, but a rich yellowish
tinge, that has a speciality about it that we call
" creamy."

Starch is sometimes said to be used for stiffen-
ing cream. It may be detected by making a
little solution in warm water, and if on adding a
few drops of iodine tincture the colour should
turn blue, starch is present; if there is no change
in colour, starch is of course absent.

Boracic acid is sometimes added to cream to
keep it from going sour. Most persons do not
look on this as adulteration. In putting aside the
milk for cream we should remember—

1. Milk at a falling temperature from 45° F.
to 40° F. raises the cream most rapidly.

2. At that temperature the volume of cream is
greater.

3. The yield of butter is more considerable.

4. The skim-milk, butter, and cheese are of a better quality.

In Fig. 21 we give the microscopic appearance of the fatty particles in genuine cream. Artificial cream is said to be made with albumen and cream faintly coloured with annatto.

Butter.—This is derived from the solid fat of milk, and besides this it should contain nothing

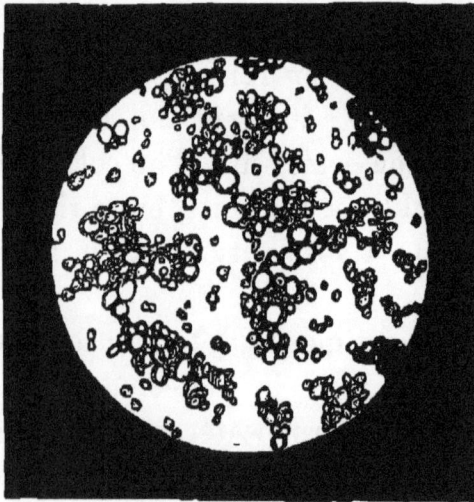

Fig. 21.—Genuine Cream.

else, unless the butter is required to keep, then a little salt should be added.

To secure good butter the fat should be well washed with cold water, so that every trace of buttermilk is taken away, or the butter is likely to go bad.

Butter is often adulterated with other fats, especially with a substance called margarine. Margarine is not bad in itself, but should not be

passed off as butter. Butter is most easily
digested of all the fats; margarine is not so
easily digested. To detect margarine in a sample
of butter is not very easy; if you can arrange
your experiments so as to fix on the temperature
at which the fat melts, this is a great help towards
its detection.

Margarine melts at 88° F., but genuine butter
does not melt till 95° F.; other fats, like lard, do
not melt till a higher temperature still is reached.
In melting a sample in a test-tube by means of a
hot-water bath, should it remain solid at a higher
point than 95° F., some other fat than margarine
is used as an adulterant.

To ascertain the melting point, this plan may
be adopted : place a sample of butter in a small
test-tube, let it be held in a vessel of water,
placed over a lighted lamp; in this vessel suspend
a thermometer. A small beaker of water standing
in a tin saucer containing a layer of sand is a
good contrivance for the outer vessel, you can
then see through the water immediately the fat in
the test-tube begins to melt. As the temperature
rises, carefully watch the thermometer, especially
after the temperature 85° F. is reached; should
the temperature go on rising to 95° F. before you
can see through the liquid yellow fat in the tube,
you have no margarine present; should it run to
fat before reaching this temperature, margarine is
present. If your specimen is pure butter, lift it
out of the bath when at 95° F., and on holding it

up to the light you have a nearly transparent
liquid of a bright yellow colour.

You may say that this is a troublesome process;
so it is, for it requires great care to detect the
exact temperature at which the fat melts; it is,
however, a most satisfactory test. To detect this
melting point some analysts use a little weighted
glass float, which sinks into the fat when it
becomes liquid. The temperature is noted immediately the weight sinks.

The method may be rendered a trifle easier
perhaps by reducing the size of the test-tube, so
that the column of melted fat to look through is
exceedingly small. The melting point is the best
indicator of pure butter, so that if rather troublesome, it is satisfactory as to the result.

We give the melting points of a few of the fats
used in adulterating butter; compare these, and
you have an indication of the fat you are
examining.

Melting points of—

Margarine	31·3° C.	88·3° F.
Cocoa-butter	34·9°	94·8°
Butter (true)	35·8°	96·4°
Beef dripping	43·8°	110·8°
Veal „	47·7°	117·8°
Mixed „	42·0°	107·6°
Lard	42·0°	107·6°
Tallow	53·0°	127·4°

The microscope will be a help in the case of
butter. Melt a little butter, drop it on to cold

water : you will get a thin layer of butter fat, which can be examined with the microscope. If it is pure you will get a uniform pattern; if margarine be present, you will have a number of small patterns. In Fig. 22 we give the microscopic appearance of pure butter. Of course if starch is used in such adulteration the microscope will reveal it at once.

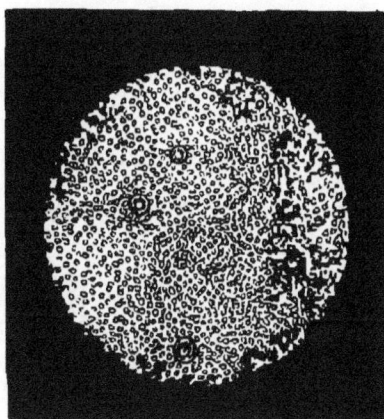

Fig. 22.—Pure Butter.

There is one other plan, which is perhaps simpler, *i. e.* take a small piece of butter, draw through it a piece of darning-cotton, so that you have a miniature candle. Set light to the cotton, let it burn for a very short time, then blow it out. If there is no disagreeable smell left behind, your sample is probably good pure butter; if, however, there is a smell of tallow, or any such disagreeable odour, you may be sure your sample is not a pure butter,

Butters that are brought to this country must not be condemned because of having lost their colour, for sometimes the butter in the centre of a tub will be quite white, and yet be a pure butter.

Frozen butters are brought over here from Australia now, and work up exceedingly well. Of imported butters, the Danish is generally the best to buy.

If butters have too much salt this must be considered adulteration. Fresh butters should only contain a little more than 1 per cent., and salt butter about 5 per cent.; any excess beyond this is not needed. Salt can be detected readily with the microscope.

Margarine.—Pure margarine is a very wholesome fat, and for cooking purposes, such as making pastry, it is to be preferred to much of that which goes by the name of "cooking butter." It is made from refined "butchers' fat," worked up with oil and milk. To make it saleable it is necessary to colour it so as to look like butter.

In its turn it comes in for a share of adulteration with other inferior fats. In Fig. 23 we give the microscopic appearance of pure margarine. As we have mentioned, margarine is not so digestible as good butter fat, but it is better to eat with bread than a badly made butter. It is perfectly wholesome, and its flavour is often very good, but it is less delicate in its flavour than butter. Butter to the palate is the most delicate

as well as the most digestible fat among our food
substances.

Cheese.—Cheese is made from the solids in
milk, including sometimes all its fat, and sometimes merely a part, often a very small part, of the
fat. Some of the richer cheeses are made of
cream added to new milk; the curds mix with the
globules of fat, all being put into the press

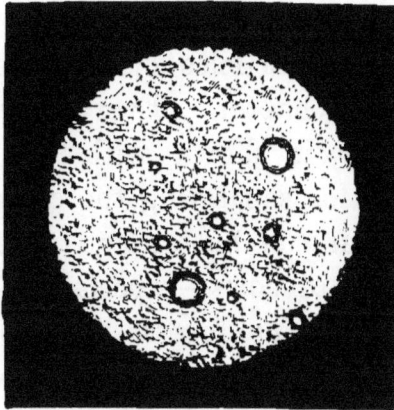

Fig. 23.—Pure Margarine.

together; the mass moulds into a solid lump, from
which all the liquid soon drains away.

There is not much adulteration in cheese,
though each make has its own characteristics.
Some of the poorer cheeses are made from skim-
milk only; the better class vary according to the
quantity of fat put into them, *e. g.* a stilton or
cheddar is very much richer than a Wiltshire or
Suffolk cheese. We mention this, because taking
fat from milk used in cheese-making does not
come under the term adulteration, but constitutes

the character of the cheese. Various ferments are added to ripen cheeses, and to give them special flavours.

If very thin slices of cheese are examined by means of the microscope, you will in most cases be able to detect it, if any foreign substance is present.

The yearly exhibition of the Dairy Farmers' Association, which takes place at the Agricultural Hall, forms a splendid object-lesson in all that appertains to the articles of food we have discussed in this chapter. Everything connected with the dairy is shown with all the latest improvements for separating the cream from the milk, the making of butter and cheese. There is also a show of cows—especially those which are the best milk-producers. It also points out the fact, that while so many wheat-growing farmers have been ruined, dairy-farmers have not only floated, but been able to make profits. It also suggests the question, Why should not we in England have more land devoted to dairy-farming, seeing that the yearly imports of butter, cheese, condensed milk, and eggs amount to several million pounds sterling? A large portion of this produce comes, it is true, from our own colonies; but for butter, Denmark alone sends us £5,280,000 worth. This is rather more than half the total quantity imported.

The children of this country would certainly be benefited by a better milk supply, for it furnishes them with every food requisite for making material for their healthy growth. Some children— London children in particular—do not have nearly as much milk as they should. According to various returns for the United Kingdom, the total yearly consumption represents about sixteen gallons of milk per head: this is only about one-third of a pint per day. Suppose we take London alone, where it seems that children want it most, the average falls to six gallons per head for the year. Among the poorer classes it would be a great improvement if parents spent more on milk and less on alcoholic beverages. The whole family would be better for the change, especially the children. We hope that the results of Technical Education in the production of dairy produce will have the effect of increasing the quantity and improving the quality, and cut out altogether every species of adulteration in this class of food.

CHAPTER V

TEA AND COFFEE

Tea.—ALTHOUGH we have two kinds of tea recognized in the market, viz. black and green, we must bear in mind that they both come from the same plant. We ought also to know that the active substance for which tea is drunk only exists in this plant, which is said to belong to China, and in one other, that is a native of Paraguay, which is called Paraguay tea.

This active substance is theine, and belongs to the same class of alkaloids as quinine. In its pure state it is a white, needle-like crystal. You will be inclined to ask, how it is we have black and green tea if they both come off the same plant? The difference is entirely due to the time of gathering the leaf and the manner of its preparation. To learn what kind of a leaf it is look at Fig. 24, where you have several represented on the stalk, and in Fig. 25 a plain ordinary leaf spread out flat. The flowers of the tea-plant are white, something like our wild dog-rose, but smaller,

being only about an inch across. The leaf is
rather fleshy, and seldom exceeds two inches in
length by one inch broad; its edge is serrated
very regularly nearly to the stalk, the veins run
almost parallel to each other from the midrib;

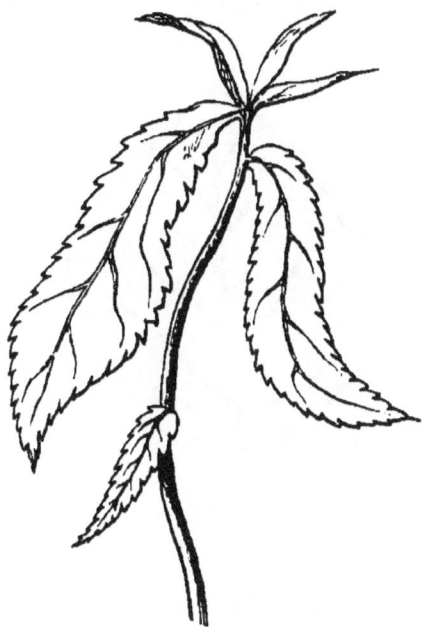

Fig. 24.—Sprig of Tea-leaves.   Fig. 25.—Single Tea-leaf
(under-side); nat. size.

before the border of the leaf is reached these ribs
turn inwards, leaving the border clear. To see
whether you have tea-leaves or not, pour a little
warm water on to a small quantity of tea, so as to
soften the leaf, then take pains to spread it out
with the back of the leaf upwards, look at it with

the magnifying glass, and compare it with Fig. 25.
The leaves are of different sizes, and some are
much broken, but you will readily see the shape
of the leaf and the arrangement of the veins.
Various leaves have been substituted for tea-
leaves, although this is not done to any great
extent. In Figs. 26 and 27 we give the forms of
some of the leaves that are occasionally substituted
for tea.

Fig. 26.—Rose-leaf.   Fig. 27.—Beech-leaf.
(Frequently substituted for Tea.)

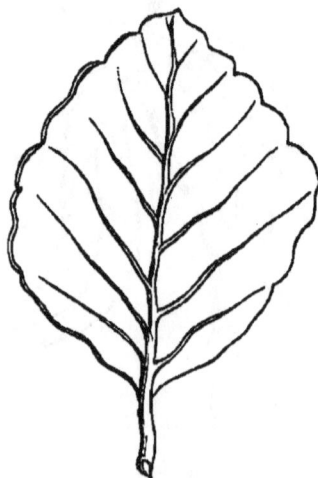

The chief adulterant is due to spent leaves—
i.e. leaves which have already been used for
"making tea;" these are collected, dried, and
bought up again.

The most recent "enormous" case is that
reported in the London daily papers of Monday,
October 22, 1894, which report detailed how the
discovery was made, the plan adopted to get such

rubbish into the market, and the cheap rate at which such tea can be sold, and the amount of fine really inflicted with the amount of fine that law allows to be inflicted. These spent leaves may be mixed with inferior or with good teas, and the quality of the mixture will regulate the price at which it comes into the market as "cheap tea."

Sometimes leaves are coloured; this colour is generally mixed with a little weak gum-water. If leaves look very green or very black colouring matter may be suspected. Take a little of the tea—a pinch—rub it between your fingers, which must be dry, drop the powder into a wine-glass of clean cold water, stir it about, and if the water is coloured, the colour comes from the surface of the tea. If there is much colour, add to the water a little potassium ferro-cyanide, the solution will then turn to a bright blue, like Prussian blue. This shows that an iron salt has been used for coating the outside of the leaf. If a black powder comes off into the water, that is black-lead, which is often used for spent leaves. Sometimes rubbing dry tea between folds of white calico takes off the colour if the tea is faced with these powders.

To detect the spent leaves, take a very small quantity of tea, pour boiling water on to it, let it stand for a short time, then pour it off, and add a second supply of boiling water; see how often the operation can be repeated so as to give a

E

coloured solution. If the leaves are spent leaves, your second decoction will be very light in colour. In a good and genuine sample of tea this operation may be repeated several times.

There are various methods of detecting adulterations. We do not know of a simpler or better than that adopted by tea-tasters in the tea-merchants' offices. They take as much tea as will balance a sixpence, put it in a cup, cover it with boiling water, let it stand for five or six minutes; the colouring matter will go either to the bottom or come to the top. The strength and aroma of the infusion is a true representative of the quality of the tea.

Some recommend that an easy way to be sure of the quality of tea is to take the specific gravity of the infusion. The solution is made by taking one part of dry tea and ten parts of water, quickly raised to the boiling point and then filtered; genuine teas will give a solution of between 1010 and 1014, taking water at 1000. This test can be applied with the bottle we recommended in Chapter I. We give the average results of some experiments in this method of testing teas.

From ordinary Congou,	Sp. gr. =		1009·88
Hyson	„	„	1013·67
Gunpowder	„	„	1012·77
With Indian teas:—			
Congou	„	„	1012·68
Pekoe	„	„	1014·32
Hyson	„	„	1013·80

We add another hint on removing the facing of

teas. Take a small sample, cover it with cold water, shake it, and pour the liquid off quickly. Notice if any sand goes to the bottom. The quantity of theine is about 3 per cent. in good teas. From a series of experiments we note that 100 grains of the following teas yielded—Congou $2\frac{3}{4}$ grains, fine Congou $3\frac{1}{10}$ grains, gunpowder $2\frac{3}{4}$ grains, Assam $3\frac{1}{2}$ grains of theine respectively.

The most satisfactory plan is to take the weight of the ash of the dry tea, which rarely exceeds 5·5 to 6 per cent. if the sample is genuine, but this process is too troublesome to give at length here.

Another plan is to examine sections of tea-leaves by means of the microscope, but we think the methods we have already given will suffice for our purpose.

We may be allowed to add a caution about making tea. Do not use water that has been boiling for some time, but use water just on the boil. Do not let it draw too long; five or six minutes is ample time, if you wish to secure the delicate aroma of the tea.

Very pure tea is now made up into tabloids; these are exceedingly convenient if a cup of this beverage is needed in a hurry.

Do not buy the cheapest teas, nor dusty teas.

Coffee.—In taking coffee we use the berry and not the leaf. The aromatic qualities of the coffee-berry are developed in the roasting, as those of the tea-leaf are in the drying. We advise all householders who use coffee to buy it in the fresh-roasted berry and not in the ground condition; not that the berries are entirely free from adulteration, but they are less liable to it. Coffee contains caffeine, that is, a white, needle-formed crystal, like theine; in fact, caffeine and theine are the same substances. Chemists tell us that the substances used in adulterating coffee are numerous, and include lupine seed, beet-root, parsnips, beans, dandelion-root, burnt malt, date stones, and acorns, but that chicory is the most extensively used, and, as far as our experience goes, it is the only one we need to look for. If you have the roasted coffee-berry, take a sample out of a well-mixed packet, spread them out, see that they all have the character of the berry; if they are broken, or are soft, examine them. You can recognize coffee-berries by biting or crushing them. Chicory, beans, and parsnips roasted and cut up about the size of the coffee-berry, cannot be passed off as such if care is taken in the examination; besides, they are always somewhat moist and sticky. At a mere glance they may be passed off as coffee.

When bought as powder, then advantage is taken of the fact that ground chicory and coffee look very much alike—or chicory finely powdered coats the ground coffee. Chicory can easily be detected by the smell. There is a sickly sweet smell in the mixture, but a clean aromatic smell with pure coffee. Take a pinch of the mixture, drop it into a glass of cold water ; if the powder floats, and the water is but slightly coloured, no chicory is present; if some of the powder falls directly to the bottom of the glass, giving the water a reddish muddy appearance, you may be sure that chicory is present: the coffee floats, the chicory sinks.   Coffee gives very little colour to cold water during the first fifteen minutes.   To

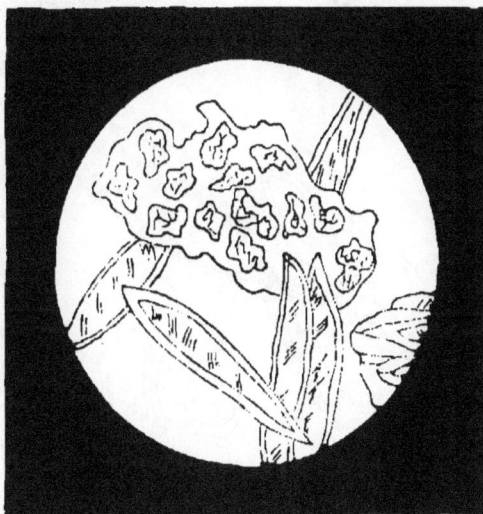

Fig. 28.—Genuine Coffee.

know the extent to which chicory is present, it is best to take a sample and look at it under the microscope.   In Fig. 28 we give a microscopic sketch of genuine coffee, and in Fig. 29 of genuine chicory, and in Fig. 30 coffee and chicory mixed.

To get a standard for examining these, take a coffee-berry, break it up in a mortar—not in a mill, unless it has been kept for coffee grinding only—drop the powder into water; see how it acts, for all coffee should act alike. The powder is slightly greasy—that prevents it from being immediately wetted. Take a little pure chicory, drop some into water in the same way; notice how quickly the water becomes coloured. Chicory contains a good deal of sugar—coffee contains very

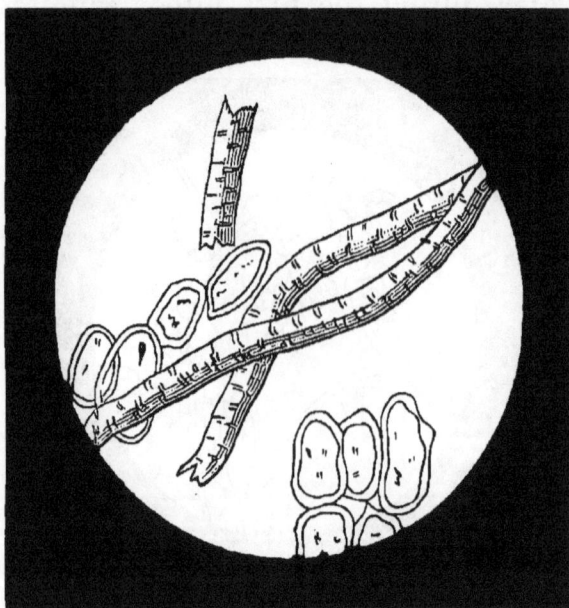

Fig. 29.—Genuine Chicory.

little. It must, however, be remembered that chicory contains no theine whatever, and therefore is of no value compared with coffee. Some people like the flavour of coffee better when a little chicory is present. The better way is therefore to buy whole coffee by itself and chicory by itself, and mix them just before using.

Examine a little pure coffee powder, and also a

little chicory powder, separately, under the microscope, and get familiar with their appearances, so
that you can readily recognize them. None of
the other substances we have named as used
for adulterating coffee contain theine. Chicory
is sometimes adulterated with other things, but
as it is so cheap, we do not think this is largely

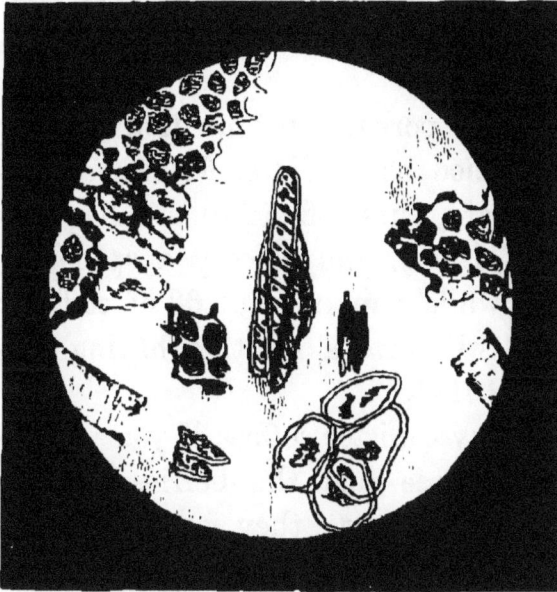

Fig. 30.—Coffee and Chicory.

done. The best coffee costs from 1s. 8d. to 2s. per
lb., the best Yorkshire chicory, 4d. per lb.

It is quite legal for a tradesman to sell chicory
mixed with coffee if he labels the packet as a
mixture of the two, and the proportion of the
mixture. The extent to which chicory is used
varies from 40 to 80 per cent., and some samples
have been found to contain as much as 90 per

cent. Some persons are so fond of chicory, that we have heard the story more than once of a mother who sent her son for "an ounce of coffee all chicory."

The merchant's test is much the same for coffee as for tea, only the quantity of coffee used for a cup of solution is as much as balance a shilling. Treat it in the same way as for tea, and if it comes out right you need try nothing further. It is easy and truthful.

We ought perhaps to mention the specific gravity method, as we did so for tea. Taking the same standard, pure coffee infusion has a specific gravity of $1009\frac{1}{2}$, and chicory a specific gravity of $1021\frac{3}{4}$, while a mixture of 60 of coffee and 40 of chicory will give a solution having a specific gravity of $1014\frac{1}{2}$.

Do not buy canister coffees if you wish for real good coffee; see that the berries you buy are not broken; see that they have been recently roasted, or warm them upon a metal plate just before grinding. Beware of coffee essences.

CHAPTER VI

COCOA AND CHOCOLATE

Cocoa.—THE active principle in cocoa, corresponding with theine in tea and coffee, is theobromine. It differs from theine in appearance, not being crystalline.

The nut is so rich in fat—more than half the nut is fat, which goes under the name of cocoa-butter; to tone down this, starch and sugar were formerly largely used, making soluble and other forms of cocoa. Since, however, the process has been adopted of making cocoa-extracts, by pressing out a large proportion of the fat, this need has not existed. In the soluble extracts of Fry's, Cadbury's, Van Houten's, Schweitzer's, and other makers, the fat has been reduced in quantity, and the cocoa has by special treatment become more soluble.

"Cocoa nibs" is the simplest and most genuine preparation of this nut, and should simply consist of the roasted seeds, crushed—not to powder—then sifted away from the husk; all the husk should be removed. Chicory may be used in adulteration of cocoa nibs, and may be detected as directed for coffee adulteration.

In cocoa pastes, starch, flour, and colouring matter have been found. Red lead has been employed for this purpose. If this is suspected, dissolve a little of the paste in water, and add a

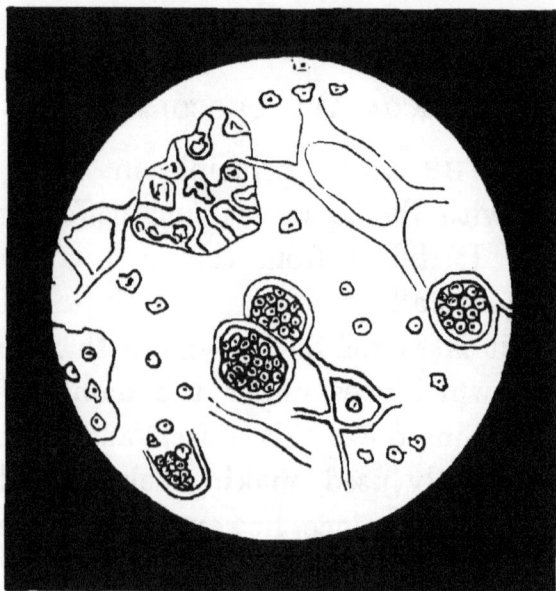

Fig. 31.—Genuine Cocoa.

little strong acid—nitric acid, for instance—then add a little solution of iodide of potassium. A beautiful yellow colouration will be at once produced. The usual test for starch may also be applied. To get an idea of the quantity of starch in any paste of this kind, the microscope must be employed. In Fig. 31 we give a sketch of genuine cocoa under the microscope.

The dry powder extracts of cocoa may be examined by the microscope as the best guide to the presence of any foreign substance. If chicory is suspected here it can be detected in the usual way. A method is adopted now by which a large portion of the fat may be extracted from the nut. This, as we have already mentioned, does away with the necessity of adding starches, arrowroot, sugar, and other substances to make the cocoa soluble and digestible. With all the natural fat present in the nut, it is much too rich for a beverage.

The cocoa-butter which is extracted is used for making the inside of the chocolate-creams and other sweets. It has the special quality of always retaining the flavour of the cocoa-nut, and never going rancid or bad. This cocoa butter at 24° C. (75° F.) has a creamy colour, and all the delicious flavour of the cocoa-nib. It is quite soluble in ether.

To detect adulteration in cocoa: the chicory can be discovered as shown in the last chapter, as when associated with coffee, while the foreign starches can be recognized by the microscope, or by their forming a jelly when mixed with boiling water. Starch can also be detected by the iodine test. The husk will be recognized by its roughness, or at once by the microscope. If the powder looks very red it may point to an adulteration from iron oxide (iron rust). Burn a little—if iron is present it will be left behind; add a few drops of sulphuric acid to the ash, then potassic ferro-cyanide, and Prussian blue will result. We do not think, however, that our readers will find this substance present.

The best security is to buy the best prepara-tions by firms who always send the best goods into the market. Among common and careless preparations of cocoa, there are still some very curious mixtures in the market, and it is to these that our remarks on adulteration are chiefly directed.

The shape of the cocoa starch granule as shown in the figure is so distinctly different to any of those we have already given; moreover, it is coloured, so that you cannot mistake it for sago, arrowroot, wheat, or potato starch, which are the substances most generally used in adulteration. Sugar may also be detected by its crystalline appearance.

Chocolate.—In the various forms of chocolate which have to be made up into soluble cakes, with some flavouring matter, starch to a small extent and sugar must not be looked upon as adulterants. To detect the chicory in chocolate or any preparation of cocoa cold water must be used, for warm water would bring the cocoa into solution.

In the mixtures of cocoa and milk the same remarks apply; the cocoa, however, must be separated, and perhaps this would be too trouble-some a process to be easily carried out.

When preparations of cocoa or chocolate are put up by good makers, who can be relied on, none of the foreign substances we have mentioned find their way into the mixtures. It is only with inferior, cheap mixtures that the presence of these need be suspected. At the same time, it is well to have at hand the means to adopt, for even if no adulteration is to be suspected, a trial exercise is an instructive exercise.

Essences and pastes of cocoa and chocolate mixtures with milk are sometimes not to be relied on. Suspected samples should be examined by the microscope. Take a portion of the paste from about the centre of the tin, make it into solution, and examine a drop of it. Repeat this with two or three drops. You should be able to see the fat of milk, cocoa, cocoa-starch, and sugar. By examining several drops you will see whether the mixture is uniform in quality, as well as convince yourself whether anything is present that ought not to be present.

Cocoa is a very nutritious beverage, which really partakes more of the nature of food than a drink. It is especially good for the young and the aged.

A peptinized cocoa is prepared specially for invalids; but, as we say, it is in the "prepared cocoas," as they are called, that we become suspicious of adulterations, where the introduction of foreign substances might not legally be called adulterations, yet the value of the cocoa would be perhaps impaired by the introduced substance.

The best form in which to buy cocoa is the concentrated cocoas of the good makers, and the cake chocolates by the same makers. They are clean, wholesome, and nutritious.

Chicory.—We have mentioned chicory as the substance used in adulterating coffee and cocoa, but we have not noticed how it is itself adulterated.

This substance grows in the form of a root very much like a parsnip, being itself the root of the wild endive. The roots are sliced and dried, then roasted, but on roasting they develop no such active principle as distinguish tea and coffee, therefore they have no refreshing principle. It is on account of its cheapness that it is mixed with coffee. Cheap as it is, however, it has been adulterated with dog biscuit, mustard seeds, roasted grain, burnt bread. These were used as a coffee substitute. Chicory seems to be a substance to which some manufacturers think they may add without stint " odds and ends " that cannot otherwise be profitably employed. That manufactured by the best makers is, however, uniformly good, and in fact may be said not to vary at all.

In detecting adulteration in some such samples as we have pointed out, the only method is to keep a good memory of what pure chicory is like under the microscope, and compare other samples with it.

Date coffee.—Some time ago a coffee called date coffee was introduced into the market, to supersede in some measure the old-fashioned beverage. It did not hold its place long. We should not have mentioned it here, but it is sometimes now used to adulterate ground coffee. The substance was made up with about one-fourth part of coffee-berry and three-fourths of roasted dates. Such coffee has too much the flavour of "all chicory," that those who are judges of coffee cannot be very well deceived by its flavour. Its presence in coffee may be detected by the methods mentioned for the detection of chicory. It acts very much like burnt sugar.

CHAPTER VII

SUGAR, HONEY, AND JAMS

WE derive our sugar from two sources, viz. the sugar-cane and beet-root. The former generally goes under the names of Demerara, Barbadoes, Jamaica, Porto Rico sugar, and is to be preferred, for with all the care that is spent in its manufacture, we never seem to get rid of all impurities in the beet-root sugar : a disagreeable smell hangs about it, especially if it is kept a long time. It is equally sweet, but we do not think it is equally nice.

Sugar is generally sold in a state of great purity, in fact some chemists make the general assertion "that sugar and bread are the only articles of food sold pure."

Pure sugar should have a nice clean crystal, clean sweet taste, and should completely dissolve in cold water. There should be no dampness, or rank smell, or a dingy-looking soiledness about it. Try samples of loaf and moist ; look at them with a

magnifying glass; dissolve portions in cold water—
no sediment should be left. Burn a specimen—
very little ash should be left behind. Sugar suffers
perhaps more than anything else by exposure to
air and dust—from the air it absorbs moisture
and dust accumulates on the surface.

Never buy a dirty sample, never buy a moist
one. Water is of course sometimes added to
increase weight. Pure sugar crystals should have
a specific gravity of 1·606, and are soluble in one-
third of their weight in cold water, and even more
so in hot water, while at boiling you get a syrupy
solution.

If you suspect sugar of having been damped
for the sake of increasing weight, take a weighed
half-ounce, put it in something, and slowly dry it
for an hour; after that time weigh it again, if it
has lost weight of course it is due to moisture;
stand it aside for another drying, then give it
another weighing; repeat the operation till it does
not alter in its weight. You can then compare
the weights at the commencement and at the finish
of the operation. The drying temperature must
not be much above the ordinary air temperature.

Treacle.—This is the uncrystallizable part of the
sugar, and only that from cane-sugar ought to be
used. That from beet-sugar is offensive both in
smell and taste. This is due to the salts and
flavouring matter which naturally belong to beet-
root syrup. These cannot be got rid of; it is,
therefore, impossible to get an agreeable smelling

F

sample, or one that can be used with comfort as
an edible accompaniment to other food.

We believe that beet-root treacle is always
condemned by the Inland Revenue, under the
Food and Drugs Act, for use in the Government
departments. Real golden syrup is good, do not
buy any other, for the inferior treacles are made

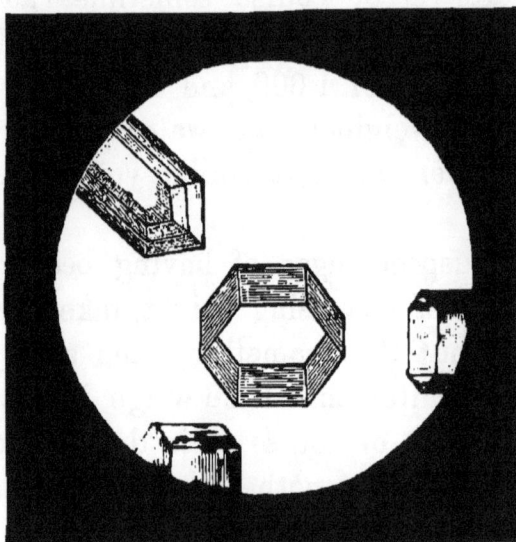

Fig. 32.—Crystals of Cane-sugar.

up of all kinds of vile material. In Fig. 32 we
give a microscopic sketch of cane-sugar.

Confectionery and sweetmeats.—These we have
now in great purity, while a few years ago we
could not have said so. Many of the stiff icings
were made of chalk, flour, plaster of Paris, and
were coloured with all sorts of poisonous substances.
The aim was to please the eye, the poisoning
qualities of the colours did not seem to be taken

into consideration. Now all this is altered, and we may certainly thank the Adulterations Act for it.

Confections and icings and children's sweets are now for the most part made of sugar only, and the bright tints are produced from various coal-tar colours, or the aniline dyes, as they are sometimes called. From the use of these you can get the brightest tints, without using sufficient to appreciably increase the weight or affect the flavour.

To test sweets of any kind—sugar sweets we refer to here, for chocolate sweets, of which so many and beautiful varieties are in the market, and which are so much "loved" by the children, are to be tested by methods already pointed out—break them up in a mortar (a small quantity only need be taken), dissolve the powder in cold water : if all is dissolved, nothing but pure sugar is present; if the solution is coloured, it is probably derived from the source we have already mentioned; should there be a sediment, filter it off, dry the powder or sediment; if it be coloured, examine it by rubbing it between the fingers, and notice whether it is tough or brittle, whether the colour is derived from the powder being stained, or whether the powder itself is the colour-producer; if so, it is probably derived from a metallic source, and should be suspected. Take a portion of the sediment, put under the microscope and examine it; this examination should confirm this point. To discover what the mineral is requires more tests than we can give here. Supposing, however, it is starch or flour, you will be able to tell at once by the microscope. Buy only the pure high-class sweets, avoid any that look to be made of "stuffs." Those made of gums and gelatines can be dissolved although not easy to pound, and can be dealt with as we have already mentioned.

We cannot help adding here a caution in reference to the cheap ices sold by various vendors who perambulate the streets of London and large towns with such sweets on barrows. These concoctions of ices are generally of the vilest kind and often injurious to health, yet they form such temptations to young folks, that the barrow of the vendor forms a favourite lounge for them.

A medical officer of health made it his special business to inquire into the composition of these "delicately-flavoured" ices. He found the compound to consist of flour, milk, eggs, and flavouring essences, and countless microbes, one deadly specimen of which is commonly found in sewage.

No wonder we often hear of children being ill, and sometimes poisoned to death after indulging in such horrible mixtures. The water in which the glasses themselves are washed is filthy, and the preparation of the "ice-cream" is carried on by individuals who have no notion of cleanly habits, or of using clean vessels in the preparation of their wares.

Honey.—The tempting little squares of honey,

"even in the honeycomb," are unfortunately
adulterated, and do not always consist of honey
pure and simple, and is frequently such as the
"busy bee" would not own as any of her make.
The cell walls are made of solid paraffin, and the
cells themselves often filled with sugar-syrup,
glucose, to which a little honey is added to carry
out the deceit a little more surely by giving it a
flavour. That sold in jars suffers in the same way.
This is one of the disadvantages of the advance-
ment of chemistry. The glucose is made by the
action of sulphuric acid on potato starch. Potato
starch again, you will say! Yes, we owe a good
deal to this humble vegetable, the potato. Who
would have discovered its virtues if Raleigh had
not?

It is difficult to find a simple test for this fraud,
except a drop of sulphuric acid—it will at once
blacken and char true bee's-wax, but will not
touch paraffin. The sugar is not the same as
cane sugar, for honey sugar is glucose, the same
substance chemically as that made from potato
starch. The microscope is the only aid we can
recommend in this case. A thorough familiarity
with the appearance of natural honey should be
of service in examining any sample you may buy.
In Fig. 33 we give a sketch of honey under the
microscope—seeds, hairs, and other items obtained
from plants will show up in such case, which in
an artificially prepared sample would be wanting.

Jams.—Here we open up a big subject, and

yet a very common one. We all know how jams
are appreciated by the children, and that they
like "real jam." We generally understand this
to mean ripe, good sound fruit, boiled down with
pure cane sugar. Such jam will retain the taste
of the fruit, and if the boiling has been carefully
done, and the jars securely covered, and they are

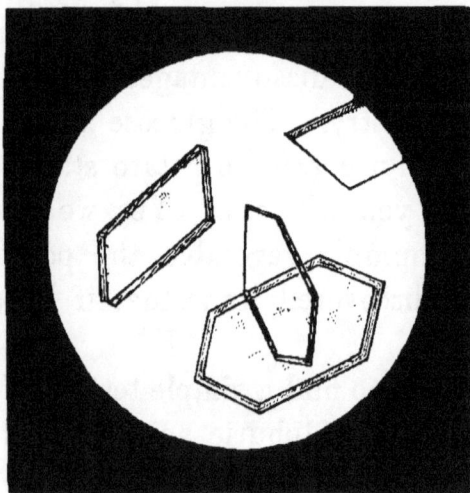

Fig. 33.—Genuine Honey.

stored in a dry place, they will remain good for
several years.

Most of the jam sent into the market is pure
and good, much more than formerly; at the
same time, if we keep to our definition that
jam should only consist of fruit and sugar, no
substance is more given to adulteration than jam.
So many tasteless vegetables, like marrows,
pumpkins, and turnips, are cut up and put with
almost any fruit, and they take the flavour. The

"mixed jams," the breakfast jams, and others, are many of them disreputable concoctions, and if they were not sweet and sticky would not be liked even by the most innocent tasters of real "raspberry." The jams of the best and purest kinds are put up in jars of largest capacities, and by firms of the highest repute, but that sold in small quantities of pennyworths and less in our poorer neighbourhoods is vile stuff.

To examine a jam, you must know what fruit to look for, and find out the cane sugar. Some very thin sections of fruit may be looked at under the microscope, and by that means cane sugar may be detected. By washing a little, however, and making a thin solution of the juice, you can perhaps tell without this; it is the pulp that must be examined, the fibres, the seeds; for "raspberry" jam, so called, is often made now-a-days without any such fruit coming into it.

The same remarks apply to marmalades. The fruit must be thoroughly looked at; portions of the pulp and rind can be examined by the microscope.

To many, jam is but a "sweet squash," for not very long ago we witnessed a transaction which told this very plainly. An individual was buying a glass jar of strawberry jam: a very good sample was shown him, in which the strawberries were whole—had not been boiled to a mash. That, however, did not suit the customer. He wanted a jam where all the fruit was so crushed that it was impossible to distinguish what fruit was employed. This kind of jam was at once supplied to him.

Buy the best jams, made by the best makers, or what is better still, if you have the convenience, "make them at home." They cost a trifle more, but you know what they consist of, and no conserve beats a good home-made jam.

CHAPTER VIII

VINEGAR, PICKLES, CONDIMENTS, AND SPICES

Vinegar.—GOOD malt vinegar is the best to employ for pickling and for general table use. Its preparation in the early processes is very much like that of brewing. The adulterations to which it is subjected, and the substitutions for it, often cause failures with pickles and other preparations requiring a good vinegar. Crude acetic acid is often used instead of vinegar in putting up common pickles, and that supplied at low-classed oyster-stalls is often but a mere concoction of dilute sulphuric acid coloured with a little burnt sugar. Pungency is no proof of a vinegar being good, but rather of the contrary—of one to be suspected. Sometimes pungency is brought about by the addition of sulphuric acid, by chillies, cayenne pepper, or grains of paradise. The latter substances are rather difficult to detect, but not so the former. Make a solution of half water and half vinegar, of two table-spoonfuls of each, add two or three drops of aniline violet: no change of colour will result if no sulphuric or other mineral acid be present. The presence of these acids to the extent of 0·2 per cent. will change the liquid to a blue tint. The presence of 0·5 per cent. of mineral acids will give a blue-green tint, and 1 per cent. of acid will change it to a bright green. This aniline violet is a very delicate test for the presence of sulphuric acid.

When there was a duty on vinegar, one part per thousand of sulphuric acid was allowed to be put with vinegar with the idea of keeping it, owing to the "ropiness," or fermentation, that is likely to take place in it, and which would decompose the acetic acid of the vinegar. Another test may be given for detecting sulphuric acid. Take a little solution of barium chloride, drop it into a mixture of the vinegar and water, it will cause a white precipitate, which if it quickly falls add more; should this precipitate be much repeated, you may be sure that more than the legal quantity of sulphuric acid has been added.

To take the specific gravity of vinegar is another good test of its quality. Taking water at 1000, vinegar should weigh 1015; anything less than this is to be suspected. Water and sulphuric acid are the principal adulterants for vinegar, and both can be detected by these methods. Avoid cheap vinegars, very pungent and very weak vinegars.

In making home-made pickles, your vegetables

will only keep well and do well by being put up
in good vinegar. Good malt vinegar, both brown
and white, may be obtained for this purpose.

Pickles.—Good pickles should contain sound,
fresh vegetables, put up with genuine spices, in
good malt vinegar. Formerly, much more than
now, metallic salts were added to give the vegetables
a fine green appearance. This to a great extent
is now discontinued. The bright green colour was
due to a copper salt. It could, however, be detected
by putting in a clean knife; it at once became
coated with metallic copper of a bright-red hue.
Pickles at present put up by good makers are
thoroughly reliable, but some put up for "immediate
use," which are sold in poor neighbourhoods by
the "pen'north," are vile enough, and sure enough
to produce indigestion. The vegetables, not the
best nor the most carefully prepared, are put into
a salt brine, lying there for perhaps two days,
brought out, then put into a mixture of crude
acid coloured and thickened by turmeric, mustard,
and starch, with a few spices, and sent out to be sold
and eaten next day. The best pickles are prepared
carefully, and not put into the market till they
are ready to be eaten.

The very inferiority of the mixture we have
described constitutes adulteration in the com-
monest kind of pickles. They cannot be eaten
with impunity. An examination of the vegetables
will enable you to detect whether all is stalk or
the inferior portions, and if they are old or

damaged; this may to some extent be discovered
on eating them.

Turmeric and substances used in making picca-
lilli may be discovered on drying up some of the
mixture, then breaking it up, and using the
microscope; any starch will at once be detected.
Flavour, crispness, and the kind of piquancy

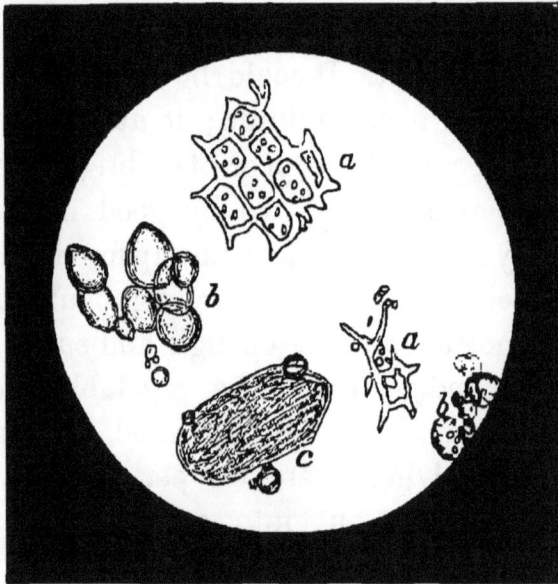

Fig. 34.—Double-superfine Mustard.

apart from pungency, distinguishes a good pickle
from a common or an adulterated article.

These remarks apply chiefly to the pickles of a
mixed character, a pickle consisting of one kind
of vegetable only—such as cabbage, onions, wal-
nuts, &c.—must be judged on similar merits. Thin
slices of vegetables can be examined by the
microscope.

Mustard.—This substance is adulterated to an enormous extent. In a loose sample you would rarely get it pure. In tins prepared by makers of repute you may depend on having it according to the guarantee on the label of each tin. A plea is often made that pure mustard does not mix well for table purposes. This is not true. It

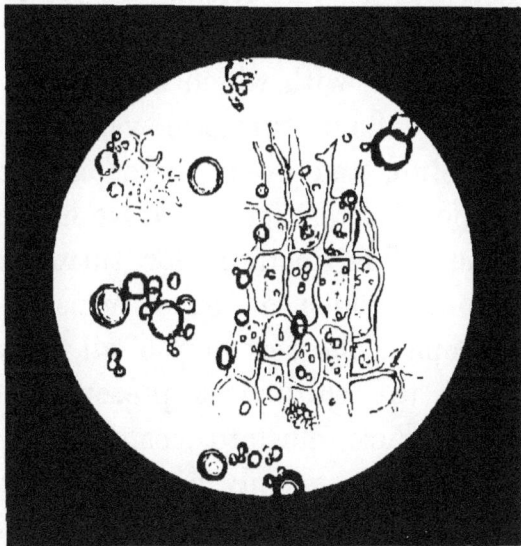

Fig. 35.—Pure Mustard.

does not keep quite so long perhaps; but that difficulty would be overcome if a less quantity were mixed at one time. Mustard is a much more enjoyable condiment when pure. In Fig. 34 we give a microscopic sketch of the so-called " double superfine " mustard. You notice that the mustard seeds, *a a*, contain oil-bearing cells; it is from these the flavour is derived. *b b* are starch granules of wheaten flour; *c c* are cells of turmeric powder. The mustard-cells look something like starch confined in skins (Fig. 35); but these husks really consist of three membranes, and are hexagonal in shape. To see this perfectly in a genuine specimen, break up a·mustard seed and examine it for your standard of comparison. If these cells are dropped into water they swell up; if you add iodine tincture to this you get no change of colour.

The substances with which mustard is largely adulterated are flour, turmeric, various starches, yellow dye stuff, and even ginger and cayenne pepper. The best test for starch we have already given is the iodine tincture; for turmeric use a little ammonia, which at once turns it brown. The microscope is the best help of all, for you can see not only the substances present, but their proportions. Pure mustard costs about 2s. per pound, its adulterants only a trifle of that cost. Mustard being frequently employed medicinally for plasters, has its strength materially interfered with when adulterated with starch to any great extent. This difficulty is got over by using mustard leaves, consisting of thin diaphragms covered with mustard flour only, which are specially prepared for this purpose.

THE COMMERCIAL VIEW OF MUSTARD.

The following paragraph from a pamphlet by Messrs. J. & J. Colman on the preparation and public appreciation of this condiment is so interesting that we quote it in full.

" Mustard of commerce or table mustard is one of those articles which has given rise to considerable discussion under the Adulteration of Food and Drugs Act, 1872. On the one hand it has been contended that pure flour of mustard only should be sold to the public. This view has been strongly advocated by prominent food analysts; while, on the other hand, men occupying equally high scientific positions see no reason to object to mixtures of mustard flour with suitable proportions of wheaten or rice flour, to moderate the somewhat coarse and bitter flavour of pure mustard, and to assist in preserving the article from decomposition, both in the dry and wet form, since perfectly pure mustard is very prone to undergo a change by variation of temperature and exposure to air, which by no means adds to its value as a condiment, in fact, renders it offensive.

" The practice adopted by the largest mustard makers, is to prepare mustards of three or four grades to suit the tastes and pockets of the public. One set of these mustards is perfectly pure, and the other set are mixed mustards, commonly called mustard condiments. Both sets are sold at precisely the same prices, and they are calculated to yield the same profits to the makers. The justification for the use of mixtures here is that the finest quality of seed is used, and by tempering this with a small proportion of wheaten flour they are able to produce a table mustard which shall possess the finest aroma and pungency, and which will keep much better, and be free from the bitterness and coarseness of flavour common to pure mustard. That the public taste approves this is manifest by the preponderance in the sale of the condiment mustard over that of the pure."

Spices.—These are generally adulterated with inferior substances belonging to the class they represent, or with fibre and absorbent substances steeped in oils imitating the flavour of genuine spice, and with partially spent samples.

Pepper.—Ground pepper often consists of dust, sand, rice-flour, sago-flour, and other substances, with a portion of genuine pepper. In addition to these, ground pepper leaves, mustard, rape seed, burnt bread, and bone-dust have been discovered. How nice a condiment is P.D. (pepper dust). The microscope helps us here again in the discovery of adulteration. We must, however, have a genuine peppercorn to begin with, and surely we can find one. Crush it up, and it should yield a powder which, on being examined with the microscope, should look like Fig. 36. You may say, " I get over this difficulty by buying my pepper whole ; " but manufactured peppercorns are put into the market, made of bran, clay, chicory, oil-cake, flavoured with pepper oil and cayenne pepper. So that the microscope is the only ready means we can recommend for the easy detection of these things.

Nutmegs, cloves, and ginger.—The former of

these are sometimes artificially produced, and flavoured with nutmeg oil. The cloves are often the spent spice dried and done up again, with some genuine oil of cloves put to them, just sufficient to pass them off, or they may be an old and damaged stock. Ginger is often made of

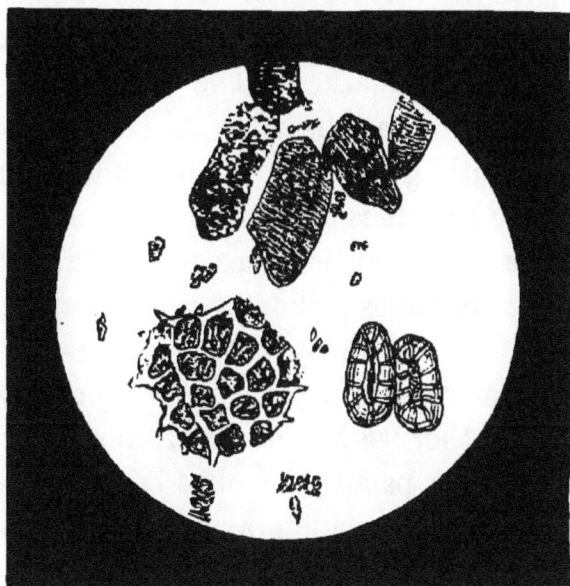

Fig. 36.—Crushed Pepper-corn.

spent samples, made pungent by various oils.

The detection of adulteration in these is troublesome, and must be tested by grating a small quantity of each, and dropping it into cold water; if it floats it may be a genuine good substance, but even then we cannot be quite sure. Only good spices should be used in making pickles.

With other spices, as with the two last, we can only recommend the microscope as the reliable

G

guide, after taking a known genuine specimen as the standard of comparison.

Olive oil.—This oil, which goes by the name of salad oil, has been so prominently before the public lately, that we are now quite convinced that very much that is sold is not the product of the olive at all. We are further informed, that not enough olives are grown to produce the amount of oil required for the market. The term salad oil is applied to it so that the seller of the sample sold under that title gets out of the difficulty of giving a guarantee that it is olive oil.

Genuine olive oil is clear, of a pale yellow tinge, and becomes solid at a temperature of about 44° F. So many other oils resemble it, that at sight we should probably be unable to distinguish whether we had a mixed sample, or even one where not a particle of olive oil was present. If we look at the specific gravity of those oils that are substituted for it, or mixed with it, they run so close together that it is not much of a guide for us. The following table gives the specific gravity at 59° F., taking water at 1·000.

Olive oil (genuine)	0·9176.
Poppy oil	0·9243.
Cotton seed oil	0·9310.
Sweet almond	0.9180.
Colza oil	0·9136.
Nut oil	0·9260.
Beech-nut oil	0·9225.

The test that seems easiest for pure olive oil, is to take one part of strong nitric acid and nine parts of oil, put them together in a test-tube, warm them till action is fairly set up, then remove the flame. Intimately mix them, then the pure oil will set in a hard mass of a pale straw colour, while all other oils will have a higher tint.

The writer of an article on "Olive-oil making near Florence" in *Good Words* for June, 1895, remarks : "Scarcely has it (the oil) left the hands of the peasants before it is manipulated and adulterated to such an extent that even in Florence pure olive-oil is almost unobtainable. The Italian Government has offered prizes for the discovery of a method of exposing the adulteration."

CHAPTER IX

BEERS, WINES, SPIRITS, AERATED WATERS, ETC.

Beers.—ENGLAND has always been famed for its beers. Before the introduction of tea into England, beer and mead seem to have been the drinks of the people. Tea and coffee are the beverages now that we like hot, beer is preferred cold.

Pure beers consist of malt and hops, and when these only are used, an agreeable, wholesome drink is the result.

It is said that beers are much adulterated, but beyond the dilution with water, we do not think this is the case. The substitutes of sugar solutions and saccharine for malt is not at present considered an adulteration, although we think it should be classed as such.

Beers, of course, vary in their strength, some contain as much as 9 per cent. of alcohol, while small beers contain as little as 2 per cent.

A sound light beer is a great desideratum, for an English workman has an idea that he must have his beer. He does his work better with than without it, it quenches thirst, and on the score of cheapness it can be sold cheaper than any other drink, unless it is cold water. A pure light beer would also do much to promote the cause of temperance.

The convictions for adulteration that have lately come under our notice have been for the addition of water and sugar, and some few months ago these cases were rather numerous. These additions lessen the tonic properties of the beer and dilute its strength. Till a standard specific gravity is imposed on beer, it is difficult to judge how far beer has been adulterated.

Salt is sometimes used in adulterating beers, and certain mixtures called "heading powders" are made use of to enable publicans to bring the beer up, to give it a "frothy head."

Salicylic acid, as we have been told, is sometimes used; this may have been added to a slight extent with the idea of preserving the beer.

Customers are in a great measure to blame for many of these adulterations—for some want a bitter beer, some a sweet beer, some a light beer, others a dark brown—all these can be supplied from the same stock, with slight additions to bring about these several results. If a beer leaves a very bitter taste in the mouth, and this taste lasts for some time, there is no doubt that the bitterness is not due to hops only, but some adulterant has been added to secure this bitterness.

The analysis of beer is rather a troublesome one, so that, with the exception of one or two tests, we shall not pursue this subject.

The "smack of age" in some beers may be due to the presence of sulphuric acid. To be sure of this, take a small quantity of the beer, add to it a little distilled water, or some water that has been well boiled. Then add a few drops of chloride of barium solution—if the mixture is cloudy it may be due to sulphuric acid, but all beers contain more or less of the sulphates of lime and magnesia derived from the water used in brewing.

To test the presence of ordinary salt, take a second quantity of the beer, add a drop or two of nitrate of silver solution. If the mixture becomes thick and cloudy salt may have been added.

The quantity of alcohol in beer may be ascertained by the process of distilling a small portion in a retort; collect the distillate, add to it enough distilled water to bring its quantity up to that of the beer in the first place used. Then take the specific gravity of the liquid. In referring this to the table in the Appendix, you will find the quantity of spirit represented by that specific gravity.

To ascertain the quantity of extract, take 100 grains by weight, in a very light saucer or porcelain dish—having weighed the saucer before using it—then evaporate the liquid, and weigh it again with all that is left in the saucer.

The best way to avoid adulteration is to buy

good quality beer from a good brewer, either in cask or bottle.

Wines.—The wines called port and sherry are made stronger or "fortified" by the addition of spirits before they come to this country at all. Many of the lower-priced wines are heightened in colour and strength by fictitious means, while some are probably mixtures of flavours and essences, coloured, to which common spirit is added, without having any acquaintance whatever with the juice of the grape. Our best wines, in fact, all bought for a fair price at respectable merchants, are fairly pure.

Home-made wines and many British wines are wholesome and refreshing beverages, and are generally free from adulteration. The colouring matter of wines is often from the juice of elderberries, blackberries, and even bilberries, but logwood, cochineal, and such substances, are only used for the mixtures of the lowest descriptions passed off as wines.

To ascertain the quantity of alcohol in wine, pursue much the same method as that recommended for beers, only that at least two-thirds of the bulk of the wine taken should be distilled over, then the quantity made up to the original bulk with distilled water. Take the specific gravity, and refer to the table for percentage of spirits. Beyond this the amateur can scarcely go, for colouring matter and essences require careful analysis to determine the source from which they come.

The light wines, like the clarets, should contain nothing but the juice of the grape; such are among the most wholesome of the alcoholic beverages. Their percentage of alcohol is low.

Spirits.—Brandy: this is a delicate spirit when pure, and is the best for medicinal purposes. It is prepared by distilling wine. Unfortunately, however, it so often consists of inferior spirits—such as may be distilled from potato mash, coloured with burnt sugar, and flavoured more or less with genuine brandy. Such a spirit may at once be detected by the harsh burning sensation produced on the palate. Various recipes for making British brandy are amusing, because they show how very little of the real article they contain.

Whisky, gin, and other spirits are largely made in England. The great adulterant is water; and another great evil exists, they are put into the market for sale and consumption when they are much too new, and thus they get the reputation of being largely adulterated. Spirits, especially when new, are contaminated with fusel oil, which has a very nauseous taste and fiery flavour, and also acts most injuriously on persons drinking such spirits. It is one of those substances that can hardly be got rid of by distillation, and only disappears by age.

The presence of this substance may be recognized by the smell. Allow a few drops of the suspected spirit to evaporate from the hand, or rub a few drops between the hands, the pungency of the smell may be taken as an indication of the quantity present.

Another test is to add a few drops of nitrate of silver to a small quantity of the spirit. Place the mixture in the sunshine : if a red tinge develops fusel oil is present, if no change in colour takes place in the mixture this substance is absent. Retailers bring down the strength of spirits considerably by the addition of water. By taking the specific gravity of the liquid you can ascertain the quantity of alcohol they contain by methods already pointed out.

Spirits are always compared with a standard called " proof-spirit," which consists of 49·37 per cent. of absolute alcohol and 50·63 per cent. of pure water. Absolute alcohol has a specific gravity of 0·7398, and proof-spirit has a specific gravity of 0·9198. In the sale of spirits they are quoted at so much " over-proof" or " under-proof." (See Table I. in Appendix.)

Rum is prepared by distilling molasses, its especial flavour being derived from an oil in the molasses. Gin derives its flavour from juniper berries, sweet flag, and other vegetable flavouring matters. It has been said that sulphuric acid is added to spirits to increase their pungency, but the flavour of this acid is too pronounced for such a purpose. The test for this is chloride of barium, as we have already pointed out (p. 90), and this can easily be applied if this substance is suspected at all.

Aerated drinks.—These are extensively used now-a-days, and are sold in strong corked and stoppered bottles, also in " syphons." Many of them are said to contain salts of various kinds, but they ought also to have a label attached telling the quantity of the salt per pint or per cent. They should contain carbonic acid gas, but many contain merely compressed air, and may be distinguished by its bubbling away rapidly, leaving the water flat to the taste. The simple aerated water—distilled if possible—is best for general purposes,[1] the various salts present being of a medicinal character. Lemonade and ginger-beer come under this heading, and are merely aerated drinks flavoured with lemon or ginger.

The various essences and substances sold for flavouring and for making sparkling summer drinks should always be used with caution, they being as a rule mere concoctions, which one is better without.

Lime-juice.—Among the most frequent drinks in summer is that from the juice of the lime-fruit. When pure this is very good—it, however, so frequently receives its sharpness from sulphuric acid. The test for this substance we have given

[1] The source of water supply for these drinks is important. It is best to have distilled water, such as that supplied by the Pure Water Company, Battersea Park, S.W.

previously. Effervescing drinks, made by mixing some acid, like tartaric or citric, with a quantity of an alkali to neutralize it, are often used in the summer-time. These should be taken sparingly, for if constantly used they are very liable to produce indigestion. The so-called "nerve tonics" also must be taken very cautiously, for many of them are more pernicious than even alcohol.

Home-made ginger-beer, and the refreshing lemonade made by slicing lemons, pouring on hot water, and adding sugar, then diluting with cold water, make agreeable and refreshing drinks, and may be taken without any fear of adulterations.

CHAPTER X

IMPORTANCE OF PURE AIR AND WHOLESOME WATER

IN our former chapters we have had it impressed on us that we should be careful to have our food as pure as we can get it. At the same time, it is even more important that the air we breathe and the water we drink should be in the fittest conditions for promoting and sustaining healthy life. These two points are so often lost sight of, that both are frequently contaminated from mere carelessness.

The air in the country, *i. e.* away from the town, is so fresh and nice, and sometimes so bracing, yet houses in these healthy spots are frequently so badly ventilated, that the rooms appear close and stuffy. We cannot here give a paragraph on ventilation, but can only say, with much emphasis, that every room in a house should have a means for the breathed air to escape, and for fresh air to find its way in. This should be the case especially with our sleeping-rooms.

A very good test, and one that is very easy to carry out, is to take a pint bottle full of water into a room, and then let the water run out, of course the air will at once replace the water. Put into this bottle one ounce of lime-water, shake it up; if the lime-water remains clear after this shaking, the air in the bottle contains less than six parts of carbonic acid to the 10,000; should any turbidity show itself, more carbonic acid is present than there ought to be. We mentioned sleeping-rooms specially, because it is here we spend nearly one-third of our lives. Do not let us put up with adulterated air, any more than with adulterated bread and butter.

Water as supplied by the great water-companies to most towns is generally sufficiently good to be above suspicion. In the summer-time, however, it is well not to be too sure of this.

A good filter is an excellent thing to have at hand. If, however, this is a charcoal filter, do not let the charcoal be in use too long without cleaning it, either by scrubbing it or by baking it. A serviceable filter can be made for a few pence. Take a flower-pot, fix a piece of sponge to block up the hole, then put a layer of magnetic oxide of iron or polarite, which can be bought for one penny per pound, cover this with a layer of sand. If in the case of very foul water, let the water first be boiled, and then run through the filter,—it will run through, practically speaking, pure.

Soft-water should not be stored in leaden-lined cisterns, and cisterns of all kinds in which we store drinking-water should be kept clean. Even in many of our model dwellings this is a point frequently over-looked. They should be securely covered ; no drain-pipe, or pipe conveying sewer-gas, should have access to water-cisterns. We frequently find people who are scrupulously clean in other matters careless of this most important one, keeping the cistern for the drinking supply thoroughly clean. Medical officers of health and sanitary inspectors' reports often tell very serious stories about the condition in which they find these water stores. In so bad a condition are some, that it is a wonder the families using them are not oftener ill than they are.

Boiled water is rather disagreeable for drinking purposes, but it is agreeably aerated if after boiling it is run through a charcoal filter. This may be effected by using another flower-pot, and covering the bottom with a thick layer of freshly-prepared charcoal.

Very hard waters should be boiled before being used, hardness in such a case being reduced by the precipitation of the lime. Hardness in water may readily be detected by its refusal to form a lather with soap. Hard-water is not therefore economical for laundry purposes.

The following are two useful tests for organic

impurities in water—impurities which should be
guarded against—

1. To a portion of the water, either in a test-
tube or in a wine-glass, put a few drops of solution
of nitrate of silver; a white precipitate will
indicate the presence of chlorides. This precipitate is readily soluble iu liquid ammonia.

2. Take half-a-pint of water, put in a few
drops of sulphuric acid, then add a small quantity
of the solution of permanganate of potash (Condy's
fluid). The solution will then be a bright pink;
should this disappear after a short time, you may
be sure the water contains some sewage matter or
other organic impurity. Drinking-water convicted
by either of these tests must be abandoned till
the evil is remedied.

Preserved and mixed food.—We cannot close
this chapter without a few words on the care
which should always be exercised when purchasing
food. In buying fresh meat, be careful that it
does not come in contact with anything that may
convey a taste to it; notice that the lean is a
healthy-looking red—no sickliness about it, that
the fat is not too white or too yellow, that there
is no flabbiness about it. In buying bacon, and
salted or dried meats, notice that everything about
it is clean, fat not yellow, no rancid smell about it,
and that by the bone it is quite sweet.

In buying sausages and mixed meats—which
are sometimes marvellous masses of mystery—be

exceedingly careful. To show what we mean, we give the analysis of a sample of sausages brought into a police-court not long ago.

The sample consisted of—

$$\left. \begin{array}{l} \text{Brown bread,} \quad\quad \frac{7}{10} \\ \text{Fat,} \quad\quad\quad\quad\quad\; \frac{2}{10} \\ \text{Flesh, meat, seasoning,} \; \frac{1}{10} \end{array} \right\} \frac{10}{10}$$

German sausage, brawn, and mixed meats of this kind, are so very liable to adulteration that it is well to know the responsible person from whom these things are supplied to be sure that they are genuine. If you take a small portion for examination, you will be able to tell whether you have meat or bread-crumbs, and whether they are good. The microscope will also help you to determine the kind of spices employed as well as the quantity. In buying tinned meats and preparations, never choose those where the ends of the tins bulge out, but always those that give a good hollow at the ends. These show that the air has been well driven out from the contents of the tin, which will be found sound and good. Never buy soiled eatables of any description.

In buying fish, freshness is the great secret of its being good, and not liable to disagree with the eater. Rely on your own judgment in selecting fish, in point of freshness,—the appearance of the eyes and gills will tell you,—and never buy any if the scales show the least phosphorescence.

In selecting vegetables, again, freshness has

much to do with their goodness; bruised and battered fruit are not good, unripe or unsound fruit should be avoided. In washing vegetables and salads, let the water be wholesome and fresh, and use plenty of it. We are just reminded of the scare which took place a little time ago, where the spread of typhoid fever was put down to the eating of water-cress which had grown on land watered with sewage-water. Had the salad been thoroughly and carefully washed, no such thing could have arisen, for whatever carried the mischief was on the outside. All such cases as this are adulteration arising from negligence in one's own home. Cleanliness is the weapon that baffles most disease.

H

APPENDIX I.

TABLE for calculating the quantity of absolute alcohol in a liquid, from taking its specific gravity at 60° F., and estimating from the same how much such a liquid is over or under proof spirit.

Proof-spirit, as given on p. 105, consists of 49·37 of pure alcohol, with 50·63 per cent. of distilled water. This mixture has a specific gravity of 0·9198, in other words it is $\frac{12}{13}$ the weight of water, *i. e.* 12 vols. of water should exactly balance 13 vols. of proof-spirit.

Specific gravity of liquid.	Percentage of alcohol.	Per cent. U. P. (under-proof).
0·9991	0·5	98·7
0·9981	1·0	97·3
0·9965	2·0	95·4
0·9947	3·0	93·3
0·9930	4·0	91·2
0·9914	5·0	89·1
0·9898	6·0	86·9
0·9884	7·0	84·6
0·9869	8·0	82·9
0·9855	9·0	80·4
0·9841	10·0	78·6
0·9815	12·0	74·1
0·9789	14·0	70·1
0·9766	16·0	66·0
0·9741	18·0	61·0
0·9716	20·0	57·0

Specific gravity of liquid.	Percentage of alcohol.	Per cent. U. P. (under-proof).
0·9652	25·0	47·0
0·9578	30·0	36·7
0·9490	35·0	26·6
0·9396	40·0	17·1
0·9292	45·0	7·7
		O. P. (over-proof).
0·9184	50·0	1·2
0·8956	60·0	18·8
0·8840	65·0	27·0
0·8721	70·0	35·2
0·8483	80·0	50·1
0·8228	90·0	63·6

In this table the specific gravity of the liquid is given to four decimal places. In the experiment with specific gravity bottle we only work to three decimal places. An average, therefore, must be taken, as must also be the case where an intermediate specific gravity is not given in the table. In using the table the distillates mentioned must only be used, for if a liquid such as a wine or a cordial spirit is taken, its specific gravity is increased by sugar and other substances present, to which the flavour of the liquid is due. Where a spirit such as whisky, gin, or a cordial has to be tested, the distillation must be conducted at a low temperature, and great care taken so as not to allow the escape of the volatile spirit. The amount of sugar and other extracts can be ascertained by taking a weighed quantity in a dish and after weighing, slowly evaporate the liquid till only the extract is left, then weigh it, and compare it with the original weight of the substance taken, and then find the percentage by ordinary arithmetic.

APPENDIX II.

In taking a temperature with a thermometer graduated according to Fahrenheit's scale, marked Fah. or simply F., it is often required to convert the same into a Centigrade reading marked Cent. or simply C., and *vice versâ*.

The following simple rules will be of assistance to those who are not already acquainted with the method of changing the readings from one to the other. In making a comparison of the readings it is necessary to remember that boiling-point, marked B.P. Fah., is 212°, and B.P. Cent. is 100°; freezing-point, marked F.P. Fab., is 32°, and F.P. Cent. is 0° or zero.

Between freezing-point and boiling-point Fah. there are 180°; between F.P. and B.P. Cent. are 100°.

Each division, called a degree, marked ° Fah. $= \frac{100}{180}$ or $\frac{5}{9}$° C., and each ° C $= \frac{9}{5}$° Fah.

Then apply the following rules.

1. To reduce a Fah. reading to C. subtract 32 and multiply the remainder by $\frac{5}{9}$. *E. g.* To reduce 104° Fah. to a C. reading—

Subtract 32, *i. e.* $104 - 32 = 72$.

Multiply by $\frac{5}{9} = \frac{72 \times 6}{9} = 40$.

∴ 40° C. is the same temperature as 104° Fah.

2. To reduce a C. reading to Fah. Multiply by $\frac{9}{5}$, then add 32. *E. g.* To reduce 75° C. to Fah.—

Multiply by $\frac{9}{5} = \frac{75 \times 9}{5} = 135$.

Add $32 = 135 + 32 = 167$.

∴ 167° Fah. is the same as 75° C.

APPENDIX III.

LIST of apparatus and chemical tests recommended in the methods for detecting the various adulterations.

APPARATUS.

Magnifying-glass.
Microscope.
Lactometer.
Creamometer.
Pestle and mortar.
Two nests of 6 test-tubes each.
250 grs. specific gravity bottle.
Chemical thermometer to 240° F.
Set of 3 beakers.
Tin saucer, 3″ in diameter.
Small set of Apothecaries' scales and weights.

TESTS.

Tincture of iodine.
Solution : Logwood.
  „   Car. ammonia.
Liquid ammonia.
Solution : Nitrate of silver.
  „   Barium chloride.
Methylated spirit.
Solution: Aniline violet.
  „   Ferrocyanide potass.
  „   Permanganate „
Sulphuric acid.
Nitric acid.
Hydrochloric acid.
Litmus test-papers, blue and red, and turmeric papers.

Readers can obtain the materials of any good druggist, and prepare all these solutions themselves from the instructions given, but for the sake of those living away from London or large towns we give the following information. Sets of tests already prepared—with or without apparatus —may be had of Messrs. Townson and Mercer, 89 Bishopsgate Street, E.C., Messrs. Orme, 65 Barbican, E.C., or of Messrs. J. & J. Griffin, Garrick Street, W.C., who will also supply microscopes. Messrs. Townson and Mercer supply an excellent Educational microscope in case for £2 2s., which is powerful enough to distinguish the various starches, while Messrs. Gregory, of 51 Strand, W.C., supply a pretty little instrument for 17s. 6d., which will just distinguish the starches, but for the more elaborate instruments the reader cannot do better than write to the firms already named, describing their wants, or to Messrs. Newton & Co., 3 Fleet Street, E.C., Messrs. Ross, 111 New Bond Street, W., or some other celebrated makers of optical instruments, who will give any information on the qualities of their various instruments, ranging in price from three guineas to one hundred guineas.

The patentee of the Tell-tale Milk-jug figured on page 50 is Mr. J. Lawrence, 56 Fulham Road, London, S.W.

Books by the same Author.

How to make Common Things. For Boys. With numerous Illustrations. Crown 8vo, cloth boards. 3s. 6d.

Simple Experiments for Science Teaching. Including Two Hundred Experiments, fully illustrating the Elementary Physics and Chemistry Division in the Evening School Continuation Code. With numerous Illustrations. Crown 8vo, cloth boards. 2s. 6d.

SOCIETY FOR PROMOTING CHRISTIAN KNOWLEDGE,
LONDON: NORTHUMBERLAND AVENUE, W.C.

PUBLICATIONS

OF THE

Society for Promoting Christian Knowledge.

THE ROMANCE OF SCIENCE.

A series of books which shows that science has for the masses as great interest as, and more edification than, the romances of the day.

Small Post 8vo, Cloth boards.

Coal, and what we get from it. By Professor RAPHAEL MELDOLA, F.R.S., F.I.C. With several Illustrations. 2s. 6d.

Colour Measurement and Mixture. By Captain W. DE W. ABNEY, C.B., R.E. With numerous Illustrations. 2s. 6d.

The Making of Flowers. By the Rev. Professor GEORGE HENSLOW, M.A., F.L.S. With several Illustrations. 2s. 6d.

The Birth and Growth of Worlds. A Lecture by Professor A. H. GREEN, M.A., F.R.S. 1s.

Soap-Bubbles, and the Forces which Mould Them. A course of Lectures by C. V. BOYS, A.R.S.M., F.R.S. With numerous Diagrams. 2s. 6d.

Spinning Tops. By Professor J. PERRY, M.E., D.Sc., F.R.S. With numerous Diagrams. 2s. 6d.

Our Secret Friends and Foes. By P. F. FRANKLAND, F.R.S. With numerous Illustrations. New Edition, 3s.

Diseases of Plants. By Professor MARSHALL WARD. With numerous Illustrations. 2s. 6d.

The Story of a Tinder-Box. By the late CHARLES MEYMOTT TIDY, M.B., M.S. With numerous Illustrations. 2s.

Time and Tide. A Romance of the Moon. By Sir ROBERT S. BALL, LL.D., With Illustrations. 2s. 6d.

NATURAL HISTORY RAMBLES.

Fcap. 8vo., with numerous Woodcuts, Cloth boards, 2s. 6d. each.

IN SEARCH OF MINERALS.
By the late D. T. ANSTEAD, M.A., F.R.S.

LAKES AND RIVERS.
By C. O. GROOM NAPIER, F.G.S.

LANE AND FIELD.
By the late Rev. J. G. WOOD, M.A., Author of "Man and his Handiwork," &c.

MOUNTAIN AND MOOR.
By J. E. TAYLOR, F.L.S., F.G.S., Editor of "Science-Gossip."

PONDS AND DITCHES.
By M. C. COOKE, M.A., LL.D.

THE SEA-SHORE.
By Professor P. MARTIN DUNCAN, M.B. (London), F.R.S.

THE WOODLANDS.
By M. C. COOKE, M.A., LL.D., Author of "Freaks and Marvels of Plant Life," &c.

UNDERGROUND.
By J. E. TAYLOR, F.L.S., F.G.S.

HEROES OF SCIENCE.

Crown 8vo.   Cloth boards, 4s. each.

ASTRONOMERS. By E. J. C. MORTON, B.A. With numerous diagrams.

BOTANISTS, ZOOLOGISTS, AND GEOLOGISTS. By Professor P. MARTIN DUNCAN, F.R.S., &c.

CHEMISTS. By M. M. PATTISON MUIR, Esq., F.R.S.E. With several diagrams.

MECHANICIANS. By T. C. LEWIS, M.A.

PHYSICISTS. By W. GARNETT, Esq., M.A.

MAPS.

MOUNTED ON CANVAS AND ROLLER, VARNISHED.

		s.	d.
EASTERN HEMISPHERE ....4 ft. 10 in. by 4 ft. 2 in.		13	0
WESTERN HEMISPHERE................... ditto.		13	0
EUROPE .... ditto.		13	0
ASIA. Scale, 138 miles to an inch...... ditto.		13	0
AFRICA .... ditto.		13	0
NORTH AMERICA. Scale, 97 m. to in. ditto.		13	0
SOUTH AMERICA. Scale, ditto. ditto.		13	0
AUSTRALASIA.. .... ditto.		13	0
AUSTRALASIA (Diocesan Map). ditto.		14	0
INDIA. Scale, 40 m. to in. 50 in. by 58 in...........................		13	0
AUSTRALIA .... ditto.		9	0
IRELAND. Scale, 8 m. to in., 2 ft. 10 in. by 3 ft. 6 in.		9	0
SCOTLAND. Scale, ditto................ ditto.		9	0
GREAT BRITAIN AND IRELAND,			
The United Kingdom of .... 6 ft. 3 in. by 7 ft. 4 in.		42	0
ENGLAND AND WALES (Photo-Relievo) 4 ft. 8 in. by 3 ft. 10 in.		13	0
ENGLAND AND WALES (Diocesan Map) 4 ft. 2 in. by 4 ft. 10 in.		16	0
BRITISH ISLES............................... 58 in. by 50 in.		13	0
HOLY LAND ....4 ft. 2 in. by 4 ft. 10 in.		13	0
SINAI (The Peninsula of), the NEGEB, and LOWER EGYPT. To illustrate the History of the Patriarchs and the Exodus..............................2 ft. 10 in. by 3 ft. 6 in.		9	0
PLACES mentioned in the ACTS and the EPISTLES. Scale, 57 miles to an inch ....3 ft. 6 in. by 2 ft. 10 in.		9	0

Photo-Relievo Maps, on Sheets, 19 in. by 14 in. :—

		s.	d.
ENGLAND AND WALES. SCOTLAND. EUROPE.			
Names of places and rivers left to be filled in by scholars.. each		0	6
With rivers and names of places......................... ,,		0	9
With names of places, and with county and country divisions in colours .... ,,		1	0
ASIA. Names of places and rivers left to be filled in by scholars... ,,		0	6
ASIA. With rivers and names of places, &c. .... ,,		0	9
NORTH LONDON. With names of places, &c............... ,,		0	6
SOUTH LONDON. With names of places, &c............... ,,		0	6
PHOTO-RELIEVO WALL MAP. ENGLAND AND WALES. 56 in. by 46 in. on canvas roller and varnished, plain 12s. coloured		13	0

ATLASES.

s. d.

HANDY GENERAL ATLAS OF THE WORLD
(The). A comprehensive series of Maps illustrating
General and Commercial Geography. With Complete
Index...*Half morocco* 42 0

BIBLE ATLAS, The. 12 Maps and Plans, with Explan-
atory Notes, Complete Index. Royal 4to.....*Cloth bds.* 14 0

A MODERN ATLAS, containing 30 Maps, with Indexes,
&c...*Cloth boards* 12 0

HANDY REFERENCE ATLAS OF THE WORLD,
with Complete Index and Geographical Statistics.
Cloth boards 7 6

STAR ATLAS (The). Translated and adapted from the
German by the Rev. E. McCLURE, M.A. New Edition,
brought up to date. With 18 Charts................ *Cloth* 7 6

STUDENT'S ATLAS (The) **OF ANCIENT AND
MODERN GEOGRAPHY**, with 48 Maps and a
copious consulting Index......................*Cloth boards* 7 6

WORLD (The), an **ATLAS**, containing 34 Coloured Maps
and Complete Index. Folded 8vo.................*Cloth gilt* 5 0

HANDY ATLAS OF THE COUNTIES OF ENGLAND.
Forty-three Coloured Maps and Index................*Cloth* 5 0

**CENTURY ATLAS AND GAZETTEER OF THE
WORLD**, containing 52 Maps and Gazetteer of 35,000
names, 4to..*Cloth* 3 6

YOUNG SCHOLAR'S ATLAS (The), containing 24
Coloured Maps and Index. Imp. 4to..................*Cloth* 2 6

**MINIATURE ATLAS AND GAZETTEER OF THE
WORLD**...*Cloth* 2 6

POCKET ATLAS OF THE WORLD (The). With
Complete Index, &c......................... *Paste grain roan* 3 6

BRITISH COLONIAL POCKET ATLAS (The), Fifty-
six Maps of the Colonies, and Index..........*Cloth boards* 2 6
Paste grain roan 3 6

PHYSICAL ATLAS FOR BEGINNERS, containing
12 Coloured Maps.................................*Paper cover* 1 0

SIXPENNY BIBLE ATLAS (The), containing 16
Coloured Maps....................................*Paper wrapper* 0 6

SHILLING QUARTO ATLAS (The), containing 24
Coloured Maps....................................*Paper wrapper* 1 0

BRITISH COLONIES (Atlas of the), containing 16
Coloured Maps.....................................*Paper cover* 0 6

THREEPENNY ATLAS (The), containing 16 Coloured
Maps. Crown 8vo.................................*Paper cover* 0 3

PENNY ATLAS (The). 13 Maps. Small 4to.....*covers* 0 1

MANUALS OF HEALTH.

Fcap. 8vo, 128 pp., Limp Cloth, price 1s. each.

HEALTH AND OCCUPATION. By Sir B. W. RICHARDSON, F.R.S., M.D.

HABITATION IN RELATION TO HEALTH (The). By F. S. B. CHAUMONT, M.D., F.R.S.

NOTES ON THE VENTILATION AND WARMING OF HOUSES, CHURCHES, SCHOOLS, AND OTHER BUILDINGS. By the late ERNEST H. JACOB, M.A., M.D. (OXON.).

ON PERSONAL CARE OF HEALTH. By the late E. A. PARKES, M.D., F.R.S.

AIR, WATER, AND DISINFECTANTS. By C. H. AIKMAN, M.A., D.Sc., F.R.S.E.

MANUALS OF ELEMENTARY SCIENCE.

Foolscap 8vo, 128 pp. with Illustrations, Limp Cloth, 1s. each.

PHYSIOLOGY. By F. LE GROS CLARKE, F.R.S., St. Thomas's Hospital.

GEOLOGY. By the Rev. T. G. BONNEY, M.A., F.G.S.

CHEMISTRY. By ALBERT J. BERNAYS.

ASTRONOMY. By W. H. CHRISTIE, M.A., F.R.S.

BOTANY. By the late Professor ROBERT BENTLEY.

ZOOLOGY. By ALFRED NEWTON, M.A., F.R.S., Professor of Zoology in the University of Cambridge. New Revised Edition.

MATTER AND MOTION. By the late J. CLERK MAXWELL, M.A., Trinity College, Cambridge.

SPECTROSCOPE (THE), AND ITS WORK. By the late RICHARD A. PROCTOR.

CRYSTALLOGRAPHY. By HENRY PALIN GURNEY, M.A., Clare College, Cambridge.

ELECTRICITY. By the late Prof. FLEEMING JENKIN

MISCELLANEOUS PUBLICATIONS.

s. d.

Animal Creation (The). A popular Introduction to Zoology. By the late THOMAS RYMER JONES, F.R.S. With 488 Woodcuts. Post 8vo. .... *Cloth boards* 7 6

Birds' Nests and Eggs. With 11 coloured plates of Eggs. Square 16mo.............................*Cloth boards* 3 0

British Birds In their Haunts. By the late Rev. C. A. JOHNS, B.A., F.L.S. With 190 engravings by Wolf and Whymper. Post 8vo. ....*Cloth boards* 6 0

Evenings at the Microscope; or, Researches among the Minuter Organs and Forms of Animal Life. By the late PHILIP H. GOSSE, F.R.S. With 112 woodcuts. Post 8vo.................................... *Cloth boards* 4 0

Fern Portfolio (The). By FRANCIS G. HEATH, Author of "Where to find Ferns," &c. With 15 plates, elaborately drawn life-size, exquisitely coloured from Nature, and accompanied with descriptive text. *Cloth boards* 8 0

Fishes, Natural History of British: their Structure, Economic Uses, and Capture by Net and Rod. By the late FRANK BUCKLAND. With numerous illustrations. Crown 8vo........................... *Cloth boards* 5 0

Flowers of the Field. By the late Rev. C. A. JOHNS, B.A., F.L.S. New edition, with an Appendix on Grasses, by C. H. JOHNS, M.A. With numerous woodcuts. Post 8vo.....................*Cloth boards* 6 0 *s. d.*

Forest Trees (The) of Great Britain. By the late
Rev. C. A. JOHNS, B.A., F.L.S. With 150 woodcuts.
Post 8vo...*Cloth boards* 5 0

Freaks and Marvels of Plant Life ; or, Curiosities
of Vegetation. By M. C. COOKE, M.A., LL.D. With
numerous illustrations. Post 8vo............*Cloth boards* 6 0

Man and his Handiwork. By the late Rev. J. G.
WOOD, Author of "Lane and Field," &c. With about
500 illustrations. Large Post 8vo. ....*Cloth boards* 7 6

Natural History of the Bible (The). By the Rev.
CANON TRISTRAM, Author of "The Land of Israel," &c.
With numerous illustrations. Crown 8vo. *Cloth boards* 5 0

Nature and her Servants ; or, Sketches of the
Animal Kingdom. By the Rev. THEODORE WOOD.
With numerous woodcuts. Large Post 8vo. *Cloth boards* 5 0

Ocean (The). By the late PHILIP H. GOSSE, F.R.S.,
Author of "Evenings at the Microscope." With 51
illustrations and woodcuts. Post 8vo...... .*Cloth boards* 3 0

Our Bird Allies. By the Rev. THEODORE WOOD.
With numerous illustrations. Fcap. 8vo...*Cloth boards* 2 6

Our Insect Allies. By the Rev. THEODORE WOOD.
With numerous illustrations. Fcap. 8vo. *Cloth boards* 2 6

Our Insect Enemies. By the Rev. THEODORE WOOD.
With numerous illustrations. Fcap. 8vo. *Cloth boards* 2 6

Our Island Continent. A Naturalist's Holiday in
Australia. By J. E. TAYLOR, F.L.S., F.G.S. With
Map. Fcap. 8vo. ....*Cloth boards* 2 6

s. d.

Our Native Songsters. By ANNE PRATT, Author of "Wild Flowers." With 72 coloured plates. 16mo. *Cloth boards* 4 0

Romance of Low Life amongst Plants. Facts and Phenomena of Cryptogamic Vegetation. By M. C. COOKE, M.A., LL.D., A.L.S. With numerous woodcuts. Large post 8vo. ....*Cloth boards* 4 0

Selborne (The Natural History of). By the REV. GILBERT WHITE. With Frontispiece, Map, and 50 woodcuts. Post 8vo. ....*Cloth boards* 2 6

Toilers in the Sea. By M. C. COOKE, M.A., LL.D. Post 8vo. With numerous illustrations.....*Cloth boards* 5 0

Vegetable Wasps and Plant Worms. By M. C. COOKE, M.A. Illustrated. Post 8vo........*Cloth boards* 5 0

Wayside Sketches. By F. EDWARD HULME, F.L.S. With numerous illustrations. Crown 8vo. *Cloth boards* 5 0

Where to find Ferns. By FRANCIS G. HEATH, Author of "The Fern Portfolio," &c. With numerous illustrations. Fcap. 8vo. ....*Cloth boards* 1 6

Wild Flowers. By ANNE PRATT, Author of "Our Native Songsters," &c. With 192 coloured plates. In two volumes. 16mo...........................*Cloth boards* 8 0

LONDON:

NORTHUMBERLAND AVENUE, CHARING CROSS, W.C.;
43, QUEEN VICTORIA STREET, E.C.

www.ingramcontent.com/pod-product-compliance
Lightning Source LLC
Chambersburg PA
CBHW021938190326
41519CB00009B/1055